And the Coastlands Wait

And the Coastlands Wait

How the Grassroots Battle
to Save Georgia's Marshlands
Was Fought—and Won

Reid W. Harris

With a foreword by Jimmy Carter
and a new afterword by Charles H. McMillan III

The University of Georgia Press
ATHENS

a Wormsloe
FOUNDATION
nature book

Publication of this book was made possible by a
generous grant from the R. Howard Dobbs Jr. Foundation.

Maps drawn by Meagan J. Duever are provided courtesy of
the University of Georgia Map and Government Information Library.

© 2020 by the University of Georgia Press
Athens, Georgia 30602
www.ugapress.org
All rights reserved
Set in 11.5/15 Garamond by Rebecca A. Norton

Most University of Georgia Press titles are
available from popular e-book vendors.

Printed digitally

Library of Congress Cataloging-in-Publication Data
Names: Harris, Reid W., author. | Carter, Jimmy, 1924– writer of foreword. |
McMillan, Charles H., III, writer of afterword.
Title: And the coastlands wait : how the grassroots battle to save Georgia's
marshlands was fought — and won / Reid W. Harris ;
with a foreword by Jimmy Carter and a new afterword by Charles H. McMillan III.
Other titles: Wormsloe Foundation nature book.
Description: Athens : The University of Georgia Press, 2020. | Series: Wormsloe
Foundation nature book | "This work was originally self-published in 2008
by Reid W. Harris"—ECIP PDF view.
Identifiers: LCCN 2019040201 | ISBN 9780820356730 (paperback) |
ISBN 9780820357201 (ebook)
Subjects: LCSH: Marshes—Protection—Georgia—History—20th century. |
Wetlands—Protection—Georgia—History—20th century. | Wetland conservation—
Law and legislation—Georgia—History—20th century.
Classification: LCC QH76.5.G4 H37 2020 | DDC 577.6809758—dc23
LC record available at https://lccn.loc.gov/2019040201

This work was originally self-published in 2008 by Reid W. Harris.

To my wife Doris,
without whose love and encouragement I would have missed the pleasures realized from my service in the Georgia General Assembly. She stood by me when many in our home county opposed what I was attempting to do.

To Fred Marland,
who provided me with a definition of the area upon which the species of marsh grasses grow, and their elevations above mean sea level. He supplied me with numerous writings describing the values of the marsh and attended many vital legislative hearings. Without his invaluable help this book would not have been finished.

He will not fail or be discouraged till he has established justice in the earth; and the coastlands wait for his law.

—ISAIAH 42:4

How still the plains of the waters be!
The tide is in his ecstasy.
The tide is at his highest height:
And it is night.

And now from the Vast of the Lord will the waters of sleep
Roll in on the souls of men,
But who will reveal to our waking ken
The forms that swim and the shapes that creep
Under the waters of sleep?
And I would I could know what swimmeth below when the tide comes in
On the length and the breadth of the marvelous marshes of Glynn.

—SIDNEY LANIER, "The Marshes of Glynn," 1878

Contents

Foreword, by Jimmy Carter   xi
Preface   xiii
Acknowledgments   xvii
Maps   xviii

I. The Threat   1
The Kerr-McGee Proposal   8
The Public Hearings   13
The Value of the Estuaries   14
The Press Reports on Kerr-McGee Proposal   16
Questions of Marsh Protection   21

II. The Legislation   23
Introduction to the Legislative Process   25
Office of Legislative Counsel   28
State Institutions and Property Committee   31
First National Bank of Brunswick   33
Hearing in the State Institutions and Property Committee   35
State Institutions and Property Subcommittee   39
House Vote on House Bill 212   42
Rules Committee   48
Second Vote by the House   49

III. The Bill in the Senate   51
Spring and Summer 1969   53
Senate Interim Committee   55
Hearing in the Senate Industry and Labor Committee   64
The Bill in the House Again   71
The Bill in the Governor's Office   76

Epilogue   81
Afterword, by Charles H. McMillan III   87
Appendix A. The Attorney General's Opinion   91
Appendix B. House Bill 212   95
About the Author   105

Foreword

My first introduction to Reid Harris was when I was a Georgia senator and he was in the Georgia House of Representatives. While his interests always have been diverse—protecting the environment, preserving the separation of powers doctrine, and providing a forum for the practice of the theater arts, to name a few—his work to protect our natural resources has been a real inspiration to me.

As one of the nation's first elected officials to express concern about the environment, Harris authored the 1968 Georgia Surface Mining Act, which protects the landscape from the ravages of strip mining. Then, facing strong opposition from many of his constituents, he fought single handedly to protect over a half million acres of one of the world's most productive areas, the Georgia Marshlands. In a truly heroic struggle, he guided the Coastal Marshlands Act of 1970 in its convoluted passage through the House and Senate. The law has stood the test of time and become a model for estuarine protection. His public service demonstrates what one person can do.

I believe Reid Harris's book not only will be instructive for those interested in our natural resources, but also captivating to anyone who reads it.

President Jimmy Carter
April 3, 2006

Preface

It would be well for the reader to understand two occurrences that transpired in 1967 prior to most of the events reported in this book. These two events involve the election of the governor of Georgia by the General Assembly and the election of the Speaker of the House by the membership of the House.

After the general election in 1967 it was determined that no candidate for governor had obtained a majority of the votes. Howard "Bo" Callaway had won the Republican Party's nomination and Lester G. Maddox was the Democratic Party's nominee. A write-in campaign conducted in behalf of Ellis Arnall, a former governor of Georgia, was responsible for the lack of a majority of the other two candidates.

At that time, according to the terms of Georgia's constitution, when no candidate for governor won a majority of votes in the general election, the election was decided by the vote of the members of the Senate and the House sitting in joint session.

A short while after the general election, but before the convening of the 1968 session of the General Assembly, as I rode from Rome, Georgia, to Atlanta, I expressed to the driver of the car, Elmer George, the head of the Georgia Municipal Association, my dissatisfaction with the fact that the Speaker of the House was chosen by the governor rather than by the vote of the members of House of Representatives. I was concerned that each Speaker, under obligation to the governor, had named and would continue to name committee chairmen and the make-up of committees at the governor's direction.

I was concerned that bills and resolutions assigned to committees by the Speaker had been and would continue to be assigned at the governor's direction, assuring "do pass" recommendations for legislation the governor favored and almost certainly "do not pass" recommendations for legislation the governor opposed, giving the executive far too much power over the legislative branch.

On arriving in Atlanta, I phoned Rep. Robin Harris (no relation), of DeKalb County. I explained my concern regarding the election of the Speaker by the membership of the House rather than having the governor appoint the Speaker. We met in my hotel room and over dinner discussed the means of having the membership of the House elect its own Speaker. We contacted a number of our colleagues in the House including Milton Jones of Columbus, Elliott Levitas of Atlanta, Roy Lambert of Madison, and Wayne Snow of Chickamauga. We held a meeting at Rep. Robin Harris's house in Decatur and worked out a plan of action.

An invitation was sent out to the House membership to meet at Stone Mountain Inn. A substantial number of representatives came and expressed the desire for the Speaker to be nominated and elected by the House membership. A month later and prior to the legislative session a caucus was held at the capital and a majority pledged to support the separation of powers doctrine and to support Reps. George L. Smith for Speaker of the House and Maddox Hale for Speaker pro tem.

In 1968 at the organizational meeting of the House, the Speaker and Speaker pro tem were elected by members of the majority party. At a joint session of the House and Senate, Lester G. Maddox was chosen governor.

During the joint session, believing that the people of Georgia should have the right to choose between Callaway and Maddox, I moved that the joint session be dissolved and that the election be given back to the people. I expressed the opinion that a governor elected by the General Assembly would be a weak governor; one chosen by the people would be much stronger. My motion lost, but many members, particularly senators, voted for the motion so that upon returning to their multicounty districts, some supporting Callaway and some Maddox, they could rightfully say that they had tried to get the election back to the people.

So, now in 1968, we had a Speaker elected by the House membership and a Senate grateful at being given the chance of returning the election to the people. These events made it possible for the House to make considerations and votes without undue pressure from the governor's office and for a grateful Senate to support legislation of a House member who had partially taken them off the hook in their votes for governor.

It is doubtful that the Coastal Marshlands Protection Act of 1970 could have passed without a House independent of the governor's power over it. Independence of the House from the governor's control and returning the separation of powers to government was a monumental step. The Senate did not pass House Bill 212 out of a spirit of gratefulness, but that spirit may have helped in securing a unanimous vote there.

Elliott H. Levitas, a Rhodes Scholar, state representative, and a five-term member of the U. S. Congress wrote to Rep. Robin Harris shortly after the securing of independence of the House no longer in the control of any governor as follows:

> I believe that a fundamental and progressive constitutional (in the broad sense) change in our state's government has occurred. This resulted from concern and conversation, from the thinking and tenacity, over the years, on the part of many, many people. In recent months, the proximate cause has been the concern of a few of us who, under your leadership, were willing to make this hope a reality and were willing to devote their personal energies to this, and without regard to their personal ambitions, rewards, or consequences. To feel that I have played a part in this long process is sufficient to make me feel that whatever service I may render, past or future, will have been worthwhile because of this alone.

Acknowledgments

The author wishes to thank three individuals in particular for their work in protecting Georgia's coastlands and for their assistance with this book:

Dr. Eugene P. Odum, the father of modern ecology, who came each time I called for help.

Jane H. Yarn, whose untiring work in spreading the news that marshland protection was needed and in spurring hundreds of individuals and organizations to contact members of the General Assembly and the governor asking for support for marshland protection.

George T. Bagby, former head of the Georgia Game and Fish Commission for his role in dissuading Governor Maddox from vetoing the Coastal Marshlands Protection Act as the governor had threatened to do.

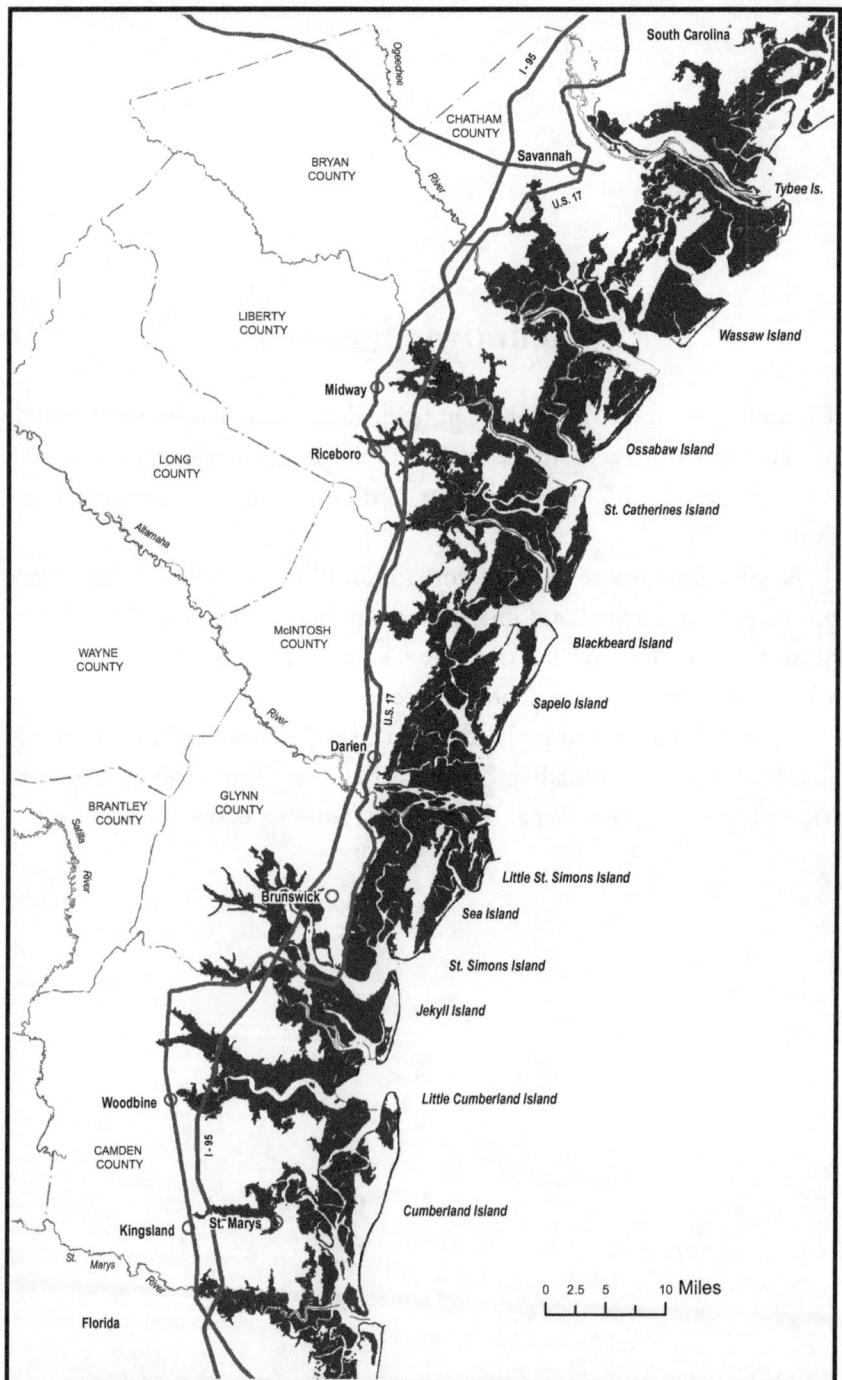

Marshlands protected under the Coastal Marshlands Protection Act of 1970. Data by the Georgia Department of Natural Resources, which estimates the state's coastal marshes and estuaries at approximately five hundred thousand acres. Map adapted by Meagan Duever, University of Georgia Map and Government Information Library, from a Georgia Department of Natural Resources "Coastal Georgia" map.

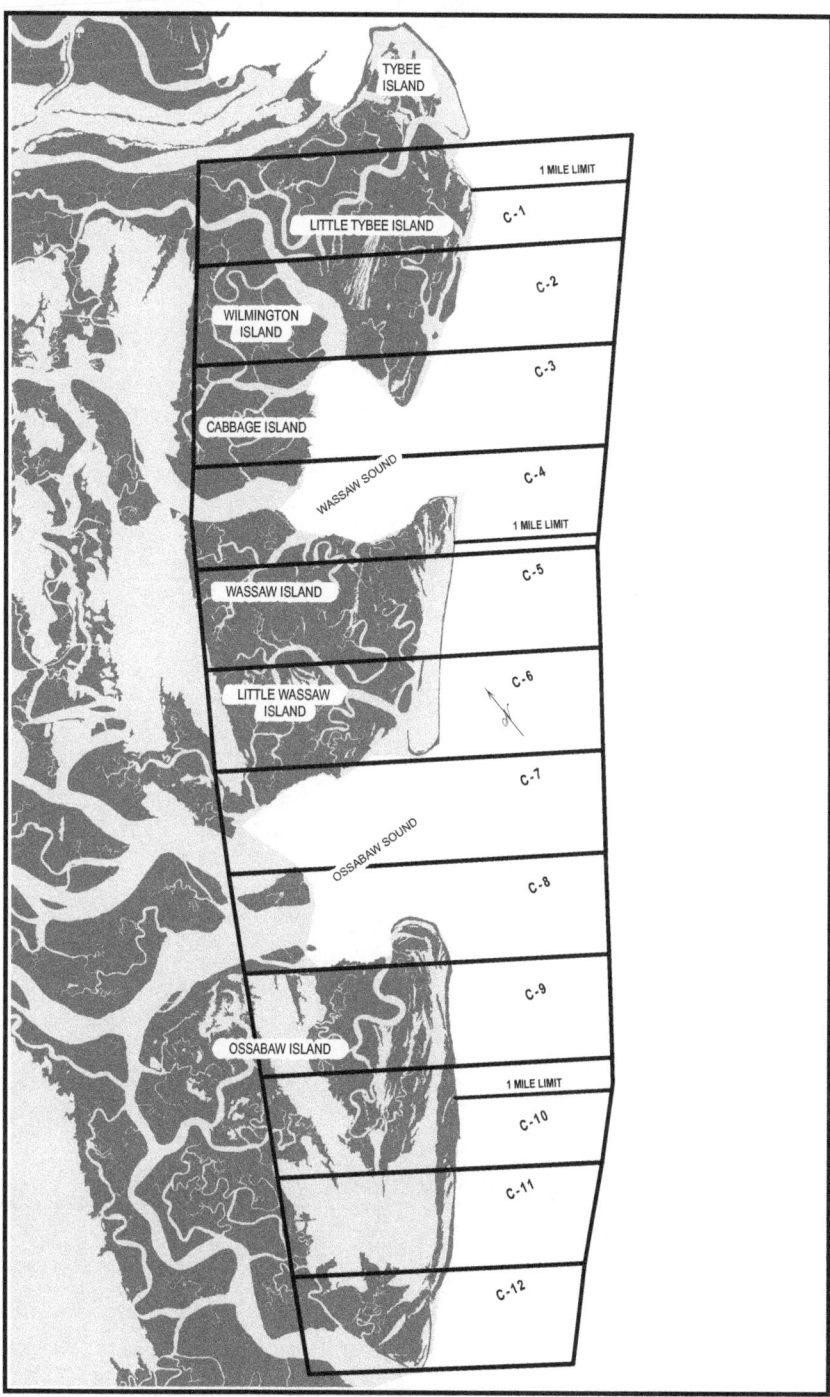

The area threatened by the Kerr-McGee Corporation's proposed lease from the State of Georgia. Map adapted by Meagan Duever, University of Georgia Map and Government Information Library, from *Georgia Fish and Game*, a publication of the Georgia Game and Fish Commission, 1968.

And the Coastlands Wait

I.

The Threat

There is always movement in the huge House Chamber at the State Capitol. During a session, legislators, visitors, and pages are always coming in and going out the high double doors at the back of the chamber. On entering, some turn immediately left and in a few steps pass through a smaller door leading to a bank of telephones and to big steel containers of cold milk. Others may turn right and move toward the large restroom.

Inside the big room, there are 196 desks, each about three feet wide. The desktops are hinged and slanted like old-fashioned school desks. Stacks of folios held together by large, shiny metal rings containing copies of bills and resolutions become thicker as the session progresses. Each desk contains a microphone and an electronic voting mechanism. Each legislator has a key to lock the voting device. Cushioned chairs on rollers stand behind each desk.

There is a wide center aisle that runs from the large double doors in back of the hall to the high rostrum at which the Speaker presides. The chamber is slanted so that the desks at the back are higher than those nearer the Speaker's dais. Two smaller aisles border the sides of the room.

On the Speaker's left the clerk and a portion of his staff are housed. From there the microphones and voting machine are controlled. The hopper into which new bills are deposited is here.

A small, slightly raised platform with a microphone stands below the Speaker's quarters. Referred to as "the well," it is from here that members explain legislation or make speeches on points of personal privilege.

The room is more than two stories high. Windows soar high above at the top of the side walls. The walls house electronic voting boards that display the names of all the members. At the back of the chamber there is a gallery that houses seats for well over a hundred spectators.

The chamber is seldom quiet. Small clusters of representatives or at times mere duos whisper to one another even as debate is taking place. A representative leaves his or her seat to make or receive a telephone call. Someone comes back. There is movement in the shuffling of newspapers placed on each desk each morning. When the stir becomes too much, the sound of the gavel restores order.

It all began in 1966. I was sitting at my desk in the legislative chamber of the House of Representatives at the State Capitol. A member of the House came up to me and handed me a small plastic bag.

"Take a look inside," he said. I opened the bag and saw a number of small pebbles.

"What's this?" I asked.

"Phosphate," he said. "I picked those up from the ground in Echols County, over in south Georgia on the Florida line."

I asked him if phosphate was being mined there.

"No, but the day will come when the mining will start."

I immediately recalled a trip I had taken that led through Florida's phosphate-mining area. I remembered the slime pits, the bleak landscape the mining left in its wake after the phosphate had been extracted, and wondered if this could be the fate of Georgia's countryside.

I kept the small bag of phosphate pebbles in my desk and from time to time opened the bag and poured some of the granules in my hand. After a short while, a few days at most, I made an appointment with a member of the Legislative Counsel's office and explained that I wanted a study resolution drawn that would provide an interim committee that would look into the need for some type of mining legislation for the state.

The resolution was drafted and placed in the hopper. It was never to see the light of day.

During the legislature's recess or early in the next session, I learned what the problem had been. A member of the House with seniority, political clout, and friendship with the Speaker had convinced the Speaker not to call the resolution for a vote.

I learned the member's name and his occupation. Before long I was sitting at a desk next to him at the front of the House Chamber. After a few generalities, I asked him about my strip-mining study resolution.

"Can't go for that, young man," he said. "Be harmful to my business."

"I thought you were in the business of selling heavy equipment," I said.

"Sell a lot to kaolin miners," he explained.

"If the mining were regulated, the miners would have to clean up their sites. It would take more, not less, equipment," I said.

"Hadn't thought of that," he said. "You going to introduce your resolution again?"

"Plan to. Are you going to fight me again?" I asked.

"Nope. Bring me a copy before you drop it in the hopper this time and I'll sign on as a cosponsor."

The study resolution was passed and a committee was formed. The committee visited phosphate mines in Florida, iron mines in Georgia and Alabama, and kaolin and marble mines in Georgia and determined that regulatory legislation was needed.

Now a question arose as to who should be named the author and sponsor of the legislation.

I remembered a story told about Sen. Richard Russell during his early days in the Senate. He wanted to sponsor a law to establish a national school lunch program for underprivileged children but was unable to get his legislation out of committee. After some thought, he went to see the senior senator from New York. The senator agreed for his name to be used and placed his name on the bill. The bill flew through the Senate and was enacted into law.

When first elected to the House, I had used three-by-five cards to help me memorize the names and faces of all the members of the General Assembly. On one side of each card I had pasted the picture of a member and on the other side the member's name, occupation and hometown.

After carefully going through the names of members, I determined to seek out a member who had served much longer than I had. I wanted someone noncontroversial who I could persuade to place his name on the mining act first.

The member agreed. The 1968 Surface Mining Land Use Act was drawn up and enacted to my great pleasure. The member I had chosen as sponsor of the legislation won a top conservation award for his effort.

The passage of the Strip Mining Act seemed to be the end of the matter until a highly significant threat came to light that would drastically affect the coastal region of Georgia. It had to do with strip mining of phosphate on the coast by an Oklahoma corporation.

After the governor signed the mining act near the end of the 1968 session, a telegram from the oldest partner of my law firm was delivered to me on the floor of the House. The wire read as follows: "Important meeting to be held this Saturday morning at 8:30 a.m."

Wondering what had prompted the wire I folded it and placed it in my suit pocket. During the day, I glanced at the message from time to time trying to determine what it might be about.

On Friday evening, I flew home from Atlanta. On Saturday morning I arrived at the law office at 8:00.

At 8:30 I walked into the eldest partner's office. The other partners and associates had taken seats, some on the sofa at one end of the room, others in chairs brought in from other offices.

The partner, whose wire had summoned me, glanced at his watch as if to indicate that I was late.

I looked at my watch, too. "Good morning," I said, glancing around the room. "My watch says 8:30."

"Take a seat," the spokesman for the firm said, puffing on his early morning cigar.

I took the only vacant chair, feeling an awkward stiffness in the air. Some had nodded at me as I entered the room, then quickly glanced away. No one had spoken a word of greeting.

Placing his cigar carefully in the ashtray on his desk and averting his eyes from mine the spokesman cleared his throat. "We've decided that you are not to run for the General Assembly again."

I was stunned. I looked around the room. All eyes were cast toward the floor.

"Do you understand what I said?"

A surge of anger hit me. "How can I not understand?"

"Well, what is your response?" His voice was cold.

"You want a response just like that?" I asked. "Not possible right now."

"When will it be possible?"

"I'll give you a reply Monday morning," I said and walked out of the room.

I quickly left the office, drove to St. Simons and out to the beach where I began to walk.

There was so much to weigh and Monday was only a day and a half away. On one side of the scales was my position as a senior partner in a fair-sized law firm, a firm of which I had been a part for ten years. There were a wife, two children, and a child on the way to support. Mortgage payments on the house.

If I stayed, what would my future relationship be with those who had met in my absence and made such a sweeping determination without any input from me?

Leaving would be a gamble. Opening a new law practice would be a difficult challenge. It might be successful; it might not be.

As I walked, watching the waves break on the shore, I knew the decision was not for me to make alone. My wife, Doris, would have to be a part of choosing the path to be taken.

And I thought about my father. He had been lifetime friends with the partner who had sent the wire and who had announced the firm's decision. I was sure he did not know what was taking place.

At home, I explained what had happened.

"You can't even think of staying with the firm," Doris said. "What will they do to you in the future?"

"I feel that way, too. But we must think of the risks of leaving."

"Risks, risks, to heck with the risks, you can make it on your own," Doris said.

Our discussion went on for some time. At the end I said that I would make two concessions to the firm. First, I would agree not to run for the assembly again after one more two-year term. Second, I would give up a portion of my income.

"It will never work," Doris said.

"We'll see."

On Monday morning, I delivered a letter to the firm setting forth my agreement.

It was flatly rejected.

The next determination was to decide when my termination would be effective.

The date was fixed at June 1.

Monday afternoon I went to see my father, deeply concerned over what opinion he would have.

The gist of his thoughts on the matter can be summarized in two sentences. "Do what you think is best. I'll send as many clients as I can to your new office."

I could not have dreamed what wonderful opportunities I had been given by my termination from the firm.

Now new office space had to be be found, furnishings and law books bought. The search for a legal secretary was a top priority.

The month of March came to a close. April sped through its course and it was soon the middle of May as I wrapped up affairs of clients from the old firm and feverishly made preparations for my new endeavor.

Then, on May 25, 1968, an announcement was made in the press that would change my life forever.

The Kerr-McGee Proposal

Kerr-McGee Corporation, a large Oklahoma-based company, made a proposal to the Georgia Mineral Leasing Commission to mine phosphate, not on high land, but in the fringes of the Atlantic Ocean, in the marsh, on the beaches, in river bottoms near Tybee, Skidaway, Wassaw, Ossabaw, and St. Catherines Islands, a huge amount of acreage covering about one-third of the coast. The proposal was made on Friday, May 24, 1968, by Franjo, Inc., a wholly owned subsidiary of Kerr-McGee Corporation which purchased Little Tybee and other islands from a Savannah antique dealer, Jim Williams, who became well known from John Berendt's book *Midnight in the Garden of Good and Evil*. The Georgia Mineral Leasing Commission postponed advertising for bids pending a study of how other states handle offshore leases.

A subcommittee of the Leasing Commission composed of Arthur K. Bolton, Georgia's attorney general, chairman; A. S. Furcron, director of the State Department of Mining and Geology; Hugh R. Papy of Savannah; and W. Perry Ballard of Atlanta was to report back to the full commission on June 20, 1968. Other members of the commission included Governor Maddox and Secretary of State Ben W. Fortson.

Officials of Kerr-McGee said that this would be a pioneering project and should produce three to five million tons of phosphate per year as related to the annual production in Florida of one million tons. The firm would dredge from 70 to 120 feet of overburden, which included the marsh, to reach the phosphate. The spoil would have to be placed somewhere as fill, presumably on the marsh, which in turn would create real estate.

There are some interesting sidelights to the meeting of the subcommittee of the Mineral Leasing Commission. It took place in Governor Lester Maddox's office at the Capitol. Maddox had invited R. S. "Rock" Howard of Water Quality and George T. Bagby of Game and Fish to attend. Bagby asked Leon Kirkland, fisheries chief, to accompany him. Maddox was enthusiastic about the possibilities. Hugh Papy of Savannah, who owned a large trucking company, thought the lease would be a boon to Savannah, and the officials of Kerr-McGee promised that tourist and recreational facilities would be built on the now-worthless but soon-to-be-filled marsh.

At the meeting, Governor Maddox, fascinated by the prospect of filling land, inquired of the attorney general if a mile of ocean were reclaimed along the Atlantic shore whether Georgia's boundary would be extended. If it could, Maddox said: "We could start digging, and we'd get the whole world if it didn't sink."

The attorney general said he didn't know if the boundary could be extended. This dialog was reported in the *Atlanta Constitution* by Duane Riner in a May 25, 1968, article.

From newspaper accounts, I learned that the attorney representing Franjo, Inc., was Charles L. Gowen. He had been the senior partner of the law firm in which I had been employed. He had served in the General Assembly of Georgia for over twenty years and my respect for him was largely responsible for my decision to run for public office. He had left the firm with which I had been associated in Brunswick to become a partner in King and Spalding in Atlanta after Griffin Bell was appointed to the United States Fifth Circuit Court of Appeals. Gowen had been the principal author of Georgia's 1945 constitution and a candidate for governor in the early 1950s.

Near the close of the meeting in the governor's office, it was agreed that at some stage a public hearing should be called regarding any proposed lease of the area to be mined and the governor stated to the press after the meeting that the commission would announce its decision within ten to twelve days of June 20 and that advertising for bids could take place in July. Immediately after the commission meeting, Leon Kirkland went back to the Game and Fish office and made a report to Jim Morrison, the commission's chief of information, and education editor of *Georgia Game and Fish Magazine*. Morrison immediately started a series of articles pointing to the dangers to the fisheries and wildlife if the proposal were to go forward. The magazine had a circulation in excess of twenty-eight thousand subscribers. Articles and editorials continued for a number of months in the magazine, many touting the need for protective legislation. Morrison also quickly contacted newsmen and editors throughout the state with news releases and suggested editorials and used his radio show, interviewing Leon Kirkland and others in broadcasts on seventy-five Georgia radio stations. Morrison and his photographer prepared a slide show on the value of Georgia's marshes to commercial and recreational fishing which Kirkland and Bagby showed Governor Maddox in an effort to persuade the Governor not to support the lease.

Rock Howard, of the Georgia Water Quality Control Board, also sounded the alarm, fearing a breach of the freshwater aquifer.

Who else joined in the forces against the Kerr-McGee proposal from the outset?

Dr. Fred Marland and Dr. Thomas Linton of the Marine Institute on Sapelo Island wrote many of their scientific colleagues on May 30, 1968:

Dear Colleague:
We have enclosed a clipping from the Savannah Morning News, reporting a potential catastrophe for the Georgia Estuarine and near shore (sic) area. Kerr-McGee will probably begin strip-mining beneath the beaches and marshes of Little Tybee, Cabbage, and Wilmington Islands. They own the first two islands and have an option on the third. The 80 to 120 feet overburden of Recent and Pleistocene marshes and beaches plus Pleistocene sediments will be deposited on the marshes. They will then mine the sounds and seaward of Ossabaw and Wassaw islands beginning at the low tide on the beaches and moving seaward to the three mile limit and beyond to fill a hole created of a fifty foot Miocene ore deposit under Little Tybee, etc. The beach front sub tidal sediments will be pumped back on the previously mined area for a real estate development. The mining of the ore and sediment will thus make a double payoff for Kerr-McGee at taxpayer's expense. Presumably, the remaining pit of 150 feet, offshore of Ossabaw and Wassaw beaches will be left open, to be filled by sediment supplied by these beaches and the others to the north (i.e. Tybee and Hilton Head is-

lands). In the Georgia Sea Islands, sediment is shifted southward along the coast in response to a southward-directed long shore current. It is hardly necessary to point out that the dredging and removal of these marshes and beaches preclude all other desirable uses such as nursery ground for fish and shellfish, aquaculture, boating, sport fishing, water skiing, water foul hunting, bird watching, etc. We believe that the landscape should be preserved, and that the greatest benefit to the citizens of this country would require that the marshes, beaches, and near shore environment be spared of strip mining. Since this Miocene phosphate extends over a wide area beneath the South Atlantic coast, a precedent will be set if this lease is granted. We are soliciting your assistance in obtaining any "ammunition" that we might use to defeat this destructive operation. We will greatly appreciate any aid that you might be able to supply, (i.e. published scientific reports) otherwise, legal aids, groups that oppose habitat destruction, or any other information that you feel may aid our cause.

Another early objection to the lease was set forth in the *Savannah Morning News*, which reported a telegram sent to Governor Maddox by Arthur Funk, a state representative from Chatham County. Funk, a Savannahian, who had retired from teaching physics, wrote as follows:

> Dear friend, may I caution you to make haste slowly in being enthusiastic about phosphate mining in our coastal counties. This is the same rock which supplies artesian water to the coastal plain and any injection of salt water could be catastrophic. Making land fill with ooze and silt dredged from the seashore or elsewhere is sheer nonsense.
> Cordially, Arthur Funk.

On Tuesday, May 28, an editorial appeared in the *Atlanta Constitution* entitled "Mining the Marshes." It seconded the governor's call for a public hearing and stated that

> it is a matter that requires the joint wisdom of experts in many fields. If it were a question of Georgia's industrial growth, we would have no reservations about the project. But this is not just a question of economics; it also involves ecology—the relationship of living things to their environment.
> The state should be sure that mining operations would not drastically alter natural life cycles.

It is interesting that Kerr-McGee's proposed sphere of operation appeared to be consistent with "zoning" of the coastal region proposed by Dr. Eugene Odum, director of the University of Georgia's Institute of Ecology. It is territory he suggested allotting to industrial, commercial, or recreational use.

By May 30, Rock Howard had written a letter to Arthur K. Bolton, the attorney general. He said that dredging operations could pierce the artesian aquifer to such an extent that salt water would flow into it thus contaminating the freshwater supply of the whole region. Howard also said that a phosphate processing operation might create a bad problem from fluoride which might be introduced as a dredging waste by-product into the shellfish breeding grounds. Citing a phosphate-mining operation in central Florida where material known as "slime waste" was accumulated in slime ponds, Howard's objections continued:

> The ponds have been broken on occasion and damaged water supplies. I am not opposed to industry, but we ought to make sure that such exploits won't leave a community in utter desolation either above or below the surface of the water.

By Monday, fishermen along the Georgia coast were voicing fears for the multimillion-dollar fishing industry. Some were contacting members of the press and letting their fears be known. They were aware that the "spoil" areas where overburden would be dumped would destroy the spawning grounds for all types of seafood.

"These folks will come, take what they want and leave. We're an established industry and want to stay," one seafood dealer said.

Time passed with little attention given in the press to the lease proposal. Then on June 14, in the *Atlanta Constitution*, an article by Phillip Gailey reported that a group of Georgia scientists and conservationists had sounded an alarm. At a meeting in Atlanta, Dr. Eugene P. Odum read a statement prepared by a group for the Georgia Conservancy.

> On the basis of information available and reviewed there is every reason to believe that extensive pollution, damage to sport and commercial fishing, reduction of potential protein food resources, and damage to recreational areas and esthetic value would result should unrestrictive strip mining be allowed over large stretches of the Georgia coastline.

Dr. Odum had many academic titles. He was named Callaway Professor Emeritus of Ecology, Alumni Foundation Distinguished Professor Emeritus of Zoology, and Director Emeritus of the Institute of Ecology at the University of Georgia in Athens. Dr. Odum helped with the establishment of the UGA Institute of Ecology, and the Marine Institute on Sapelo Island. His textbook *Fundamentals of Ecology* was one of the first published in the field of ecology and has been translated into numerous languages.

The article by Gailey reviewed the Franjo, Inc., proposal and concluded with a statement that caused the raising of eyebrows. Gailey had done his homework. He cited an article in the June issue of *Fortune* magazine which quoted Dean McGee, the chairman and chief executive officer of Kerr-McGee, as saying that, although he was constantly reevaluating potash and phosphate mining, he had no immediate intention of entering an already overcrowded field. It was clear from the article that Governor Maddox was waiting anxiously to find out how much the mineral lease would mean in terms of revenue to the state and that the officials with Franjo, Inc., wanted to go forward. The coastal press had been relatively silent on the matters surrounding the lease.

On June 16, a report in the *Atlanta Constitution* stated that the Georgia Water Control Board had joined the fight against tideland phosphate mining and had decided to send a representative to public hearings to argue against such operations. In the article by Ron Shinn, Rock Howard was quoted as follows:

> If we are going to maintain our perpetual industries—the shrimp, fish, and crab industries, we have got to protect them. You can't have your cake and eat it too. We have to be very selective as to what industries we introduce into Georgia's waters. God gave us this great perpetual resource, and we ought to take care of it.

The article further quoted Howard as saying that unless some "super sophisticated" new equipment was used instead of the usual hydraulic dredging it would be impossible to conduct phosphate mining off the Georgia coast "without destroying a very valuable natural resource."

The Public Hearings

It was reported that two public hearings were to be held regarding the proposed lease; one in Savannah on August 26, and one in Atlanta on September 16, 1968.

It was noted that the area in question to be leased encompasses about thirty miles of the Georgia coastline and contained about seventy-two thousand acres. One of the proposed ground rules for the hearings required that anyone wishing to speak must submit a written notice by registered mail to the state attorney general, Arthur K. Bolton, ten days prior to the hearing. Some of the Water Quality Control Board's members were unhappy with the rule, surmising that the rule was designed to hinder those opposed from expressing their opinions. The requirement of notice was dropped. After the warning expressed by the Georgia Water Quality Control Board and the statements of Dr. Eugene Odum, and the articles and editorials in *Georgia Game and Fish Magazine*, the coastal press began an aggressive reporting campaign.

The first page of section 2 of the *Savannah Morning News* edition of June 16, 1968, was devoted to an article entitled "Phosphate Mining—Asset or Liability?" Photographs of mining areas in Polk County, Florida, headed the article, clearly showing the devastation brought about by the strip mining of phosphate.

During the days in which the mineral lease was being discussed, I became more and more convinced that I would have to take a stand as a member of the General Assembly of Georgia. (Ecology was at the time a new word to me. It was also new to a large segment of the population.)

The Value of the Estuaries

I contacted the Marine Institute on Sapelo Island and explained that I needed help in understanding the ramifications of the proposed mineral lease. Shortly thereafter an appointment was made with marine scientists who appeared in my office in Brunswick, loaded with books, papers, and pamphlets to educate me on Georgia's marshes.

Drs. Fred Marland and John Hoyt gave unstintingly of their time and knowledge to educate me about the estuaries. They also went to Savannah and Atlanta to the hearings about the lease. They were present later in hearings before legislative committees in the House and the Senate. (Dr. Marland has remained a good friend over the years. Dr. Hoyt was killed in a glider accident shortly after the Marshlands law was passed.)

Conversations with these two men and reading the materials provided to me changed my whole perspective of the marsh environment. The marsh landscapes for me had not been, until then, a productive area. It provided vast vistas of wonderful thought-enhancing space. The marsh was to me as it had been to Sidney Lanier in his poem "The Marshes of Glynn," which I had memorized in high school.

Now I became aware that various life forms were teeming amidst the blades of grass and in the mud beneath the surrounding waters. There were organisms there so small that they could not be seen except with a microscope. Those small organisms were a part of the food chain that led to crabs, oysters, shrimp, and fish, and eventually to the mouths of man.

Drs. Marland and Hoyt taught me that the marsh provides a basin for catching the river silts that travel from inland to estuaries. They taught me that the marsh acts as a buffer in times of storms and that the area acts as a great filtering plant and that the marsh breaking apart into detritus feeds fish as far as fifty miles offshore. Because of the location of these marshes, sheltered by the out-jutting of the Florida peninsula and the eastward migration of the coast toward the north, flooded by ocean waters with exactly the right level of salinity, and a favorable temperature range, the production of this zone is the greatest on earth, producing more tonnage per acre each year than the best corn belt farms can produce per acre with no cultivation. Only reaping the harvest is required.

After these lessons, as I drove home along the four-mile causeway that leads from my home on St. Simons to my office in Brunswick, the marsh spreading to long vistas north and south had more meaning to me. This expanse of marsh is interrupted by four rivers, Frederica, Little, Mackay, and Back, and by Terry Creek. The colors of the grasses are constantly changing. In spring light green appears and through time slowly moves to darker shades of green. By late fall, tan arrives. In winter the tan, in the rays of the setting sun, becomes golden and the blades of grass die off, leaving the roots for the cycle of the coming year. In spring the grass is limp and flexible. During the summer it becomes stiffer, less pleasing to the touch. By winter the dying grass is brittle and rough, capable of harming the skin. These changes are so gradual that the alterations are perceptible only through long periods of time.

Besides the color and texture, the odor of the marsh is unique. Many who were born and raised in the coastal area, on returning home from trips inland, roll down the windows of their cars to smell the marsh's fecundity.

And what are the sounds of the marsh environment? When the tide is at its lowest, there are little popping-sloughing sounds and the smallest rivulets trickle into the rivers and creeks. When the wind is up, there is whistling through the leaves and reeds. And there are times of purest silence.

I came to understand that protection for such a productive region was essential. How to protect the marsh became a consuming question. It would take months to reach a conclusion.

I wrestled with a number of questions. Who was the English king who determined that some lands should be set aside for public use, for the fisheries and for hunting? Had the state of Georgia's General Assembly abrogated its right to enact zoning legislation in favor of cities and counties? What was the state of the ownership of the coastal marshlands? What is the true meaning of *jus privatum* (literally "private law"; in this context the proprietary right in the use and possession of lands beneath tidal waters) and *jus publicum* (literally "public law"; in this context the collective rights of the public to fully use and enjoy trust lands and waters for commerce, navigation, fishing, and other related purposes)? Had other states enacted laws to protect their coastal areas? If laws had been enacted, what did they provide and how soon could I get copies?

All of these questions and many more were haunting me as time for the first hearing on the mineral lease approached.

The times for the two hearings on the proposed lease were changed. The first was in Atlanta on September 16, 1968. There were so many who wished to express their views that it was carried into September 17. The second hearing was held in Savannah on September 30. The crowd was so large that the venue was changed from the Savannah Electric Building to a large ballroom at the Desoto Hilton Hotel. Most of the speakers expressed opposition to the lease, citing the concerns already expressed regarding the aquifer and the dangers to the fishing industry.

The Press Reports on Kerr-McGee Proposal

On August 7, 1968, prior to the two public hearings called by the Mineral Leasing Commission, A. F. T. Seale, a senior vice-president of Kerr-McGee Corporation, was interviewed by Bill Carpenter of the *Savannah Morning News*. The article was sent to me by Dr. Fred Marland, who had written in large letters along one margin "HOGWASH & SUCCOTASH!!" Marland was referring to Kerr-McGee's primary purpose of using the overlying strata as fill for development and secondarily for phosphate ore. The article reads in part as follows:

> Kerr McGee officials today said phosphate is their primary objective in seeking a tidewater mineral lease but mentioned possible development of Little Tybee and Cabbage Islands as a secondary venture.
>
> They said phosphate mining operations in the Savannah area possibly could begin by 1971–72 if their company is awarded a state mineral lease by the end of this year....
>
> Seale displayed a map of a proposed "Camelot: A Kingdom by the Sea" which he said had been developed by Franjo, Inc. prior to the company's purchase by Kerr McGee. The map was of Little Tybee Island.
>
> Although the map represented only an idea, according to Seale, it did offer a suggestion of what could be done with material covering phosphate deposits "if you've got to handle that overburden anyway."
>
> Kerr McGee purchased Cabbage Island from local antique dealer James A. Williams through its subsidiary Franjo, Inc. for $650.000 according to a deed filed Aug. 1, in the Chatham Court House.
>
> Questioned earlier this morning concerning the sale of Cabbage Island, Kerr McGee officials said the deed had not yet been filed. Officials said they had submitted the bid on the offshore lands under the Franjo name because it was an existing corporation in Georgia. However, Kerr McGee also has blending operations in Tifton.
>
> But phosphate is really what the company is after, he asserted, because of a predicted boom in the fertilizer market in the early 1970s. He contended that the market is flooded at the present time.
>
> William R. Fort, manager of the plant food manufacturing operations, said the company expects to mine at least three million tons of phosphate ore from the underwater areas per year.
>
> Any smaller amount, he explained would make mining operations impractical from an economic standpoint. He said preliminary drillings indicate the offshore ore is about 66–67 per cent bone phosphate of lime.
>
> At this percentage, however, the company would only have to mine approximately 1.9 million tons in order to meet the predicted $500,000 in royalties to the state.
>
> Seale said the mining method to be used is "out and out" dredging. The overburden would be used to fill marshland, he said, and colloidal clays would be piped out to sea and dumped.

Only marshlands owned by Kerr McGee would be affected by the mining and no state-owned marshlands would be involved, the vice-president said.

As for the possible damage to fish or shrimp breeding grounds in the area, he said only those on Kerr McGee property would be affected and the wildlife could simply find other breeding areas.

Fort admitted that the company has conducted a survey of "heavy mineral" deposits in the area and said "we are definitely going to look for them and try to find them."

He discussed the importance of possible deposits of manganese deposits in the area and said a study earlier this year by the Chief State Geologist, Jesse H. Auvil, was simply a "literature study."

As for the claim that offshore mining is not economical, geologist Fred Hohne said it is easy "to get a lot of university professors to say what phosphate is but they don't know anything about the market."

Should the lease be granted, Seale said some form of preliminary processing plant might be installed on Little Tybee to take care of separating phosphate from other sediments by washing.

Sea water could be used in all stages except the final rinse, he said, which would require fresh water. The final state, however, would not have to use potable water from the aquifer, he added.

Although he made no definite commitment, Seale said it is possible that water from the Savannah River or other sources might be used to supply the station's 1.5 million gallons a day requirement.

If granted a lease, Seale said the first year probably would be spent developing detailed engineering plans for the area and determining how much of the 25,000 acres actually contains phosphate deposits.

Any area void of phosphate can be dropped from the lease, he said, and would not be subject to the annual $3 per acre rental fee under the terms of the lease.

Although earlier mention of the building of a processing plant here had been made, Seale said it is a possibility and depends to a large extent on whether favorable power rates could be negotiated.

He said phosphate had to be processed by two methods, the electric furnace procedure or a wet acid method which utilizes sulphuric acid. The sky rocketing costs of sulphur, however, are prohibitive, he said.

If power is not available, however, then the separated phosphate will have to be transported elsewhere for further treatment or sale. This would have to be done by water, he noted.

Water travel and shipment of equipment would necessitate the dredging of channels in the Savannah Beach area, Seale said, and would open up the waterways in that area. Location of Kerr McGee here possibly could create between

300–800 jobs and open the way for other industrial expansion through chemical industries and other related concerns, Seale said.

Neither of the two processing methods mentioned would result in either air or water pollution. As for damage to the marsh, Seale said, "you've got to disrupt something for progress."

The article brought immediate disappointment to many and raised a host of questions to others. There had been reports that Kerr-McGee's activities would mean fifteen hundred jobs in the Savannah area. Now a senior official was saying something between three hundred and eight hundred would be employed. Why was this project so uncertain? Reports had it that a processing plant would be built costing some $25 million. Now it appeared that the possibility of any plant was far from certain.

Were the revenues payable by the corporation to the state of Georgia for the lease only $500,000 and the per-acre fee the mere pittance of $3 a year? There were too many "mights" and "possibilities."

For some reason this discussion centered on twenty-five thousand acres of marsh. No mention was made of the area of ocean bottoms measuring thirty miles long and out to the three-mile limit—some seventy-five thousand acres. For some reason newspaper accounts regarding the size of the area Kerr-McGee proposed to lease was stated to be twenty-five thousand acres.

How that number was arrived at has always puzzled me. A cursory glance of the map attached to the lease proposal, and shown near the start of this book, shows an area extending from the road to Tybee to the southern tip of St. Catherines Island. A third of the Georgia coast was involved. Suppose the area *were* only twenty-five thousand acres. The average depth of the overburden above the deposits was estimated to be one hundred feet. If the overburden were sliced into one-foot-thick sections and were laid side by side, the slices would cover an area of 2.5 million acres! The state of Rhode Island contains 776,950 acres. The twenty-five thousand acres sliced in one-foot sections would equal 3.21 times the area of Rhode Island. Now add the thirty miles by three miles of ocean bottom; the acreage is astronomical. How great the roiling of the waters would be—oxygen depletion rampant!

An official from Kerr-McGee stated that the firm had employed an engineering firm to determine a new methodology of extracting the phosphate, but he said the firm had not yet come up with any new proposals. (No new concept was ever produced.)

It was immediately before the last hearing that, in speaking to a member of the press, I expressed the view that legislation was needed to protect the marshes.

Shortly after the Savannah hearing a petition of injunction was brought against Governor Maddox and the other members of the Georgia Mineral Leasing Commission asking that the commission be enjoined from entering a mineral lease with Franjo, Inc. The plaintiffs were the two lawyers who brought the suit. A preliminary injunction was granted and a hearing was set for ten days after the filing of the suit. No hearings were ever held on the litigation and the suit was dismissed by the plaintiffs.

On October 13 and 14, 1968, a conference to consider the future of the coast was held at the Cloister Hotel on Sea Island, Georgia (the Conference on the Future of the Marshlands and Sea Islands of Georgia, June 1970; 128 pages). In the audience were the owners of the islands then in private ownership, or their representatives.

The panel of speakers was made up of a diversity of experts ranging from ecology, geology, archeology, to regional planning and esthetics. Among the speakers were Thomas W. Richards, president of the Nature Conservancy, Dr. Eugene P. Odum, Dr. Frederick C. Marland, and Dr. John H. Hoyt. I was asked to speak. From my statement it is evident that I was grappling with concepts regarding the type of legislation needed to protect the marsh.

In early October, a five-member group of scientists from the University System of Georgia, headed by Dr. Eugene Odum, filed a report to the Mineral Leasing Commission once more pointing out the dangers of phosphate mining in the area proposed, and at a Mineral Leasing Commission meeting on December 5, 1968, the commission returned a good-faith deposit to Franjo, Inc. At the same meeting the attorney general explained that since the commission had taken no action prior to November 30, because of the injunction, the deal was dead since the bid had expired. Did returning the check and announcing that the leasing deal was over mean that the threat was over? During a long period rumors circulated that Franjo, Inc., or people associated with Kerr-McGee Corporation were buying maps and making inquiries in coastal counties, but no one ever ascertained whether those rumors were merely rumors. As far as I was concerned, there would still be a threat until legislation was in place to protect the coastal zone resources.

Questions of Marsh Protection

During the rest of the year, I continued to wrestle with the question of what type of legislation was needed. I began to feel that zoning was not the answer. I knew that four counties bordering the estuarine area had no zoning ordinances and that the two counties which did, Glynn and Chatham, did not provide protection against filling and dredging marshes. I had learned that several states on the northern seaboard had some regulations regarding their state estuaries. I wrote the secretaries of state for Maine, Massachusetts, Delaware, Rhode Island, and Connecticut asking for copies of the legislation regarding controls. Over a period of time, replies to my inquiries were sent to me. None of these state's statutes seemed suitable to what I had in mind.

And even as I was grappling with all these questions, I was in the midst of an election to retain my seat in the General Assembly against an opponent of unknown strength. How ironic it would have been if after leaving the law firm I were defeated at the polls. Once the election was finally over, though, and my place in the assembly was secured, I turned once again to the thoughts of going ahead with the preparation of a bill to be introduced when the assembly convened in January of 1969.

I forwarded the statutes that I had gotten from other states together with some suggestions to the Office of Legislative Counsel in Atlanta. I had come to the conclusion that there should be some local input, perhaps from the commissioners of roads and revenues of the counties in which marshes to be affected by filling or dredging were located and that a state board should be created to oversee a permitting process. I felt that a board created for this process should be composed of those concerned for the environment. I wanted to have a strong initial board so that if the legislature determined that its composition should be changed, there would be a chance of retaining a majority that would consider, first and foremost, the public interest.

II.

The Legislation

Many significant changes regarding who may walk into the Capitol have been made since the convening of the General Assembly of Georgia in January 1969. At that time the Capitol building was open to all comers and on the first day of the session the size of the crowd was large and diverse. People swarmed through the corridors. Some carried placards. Others clutched brightly colored pamphlets. "Stop Capital Punishment"— "Repeal the Sales Tax." The largest number of signs read: "Save Our Children—Vote Down Fluoridation."

Teachers led groups of schoolchildren up and down and around the huge circular rotunda. The wide marble stairs teemed with boys and girls, pushing and shoving, joking and laughing as they waited before a bank of cameras that were mounted on tripods at the bottoms of the stairways. The noise permeated the building, echoing from the dome high above. On both sides of the third floor, near the entrances to the Senate and House Chambers, there was handshaking and backslapping. When hands or backs could not be reached, nods and winks were exchanged.

Professional lobbyists, briefcases at their feet, kept themselves well separated from each other. Representatives and senators making their way to their respective chambers passed a gauntlet of special interests—railroads, trucking, farming, telephones, electric power, airlines, banking, labor, and schools. Longtime assemblymen could sit in their home districts and name the lobbyists in the line. Lobbyists had their chosen spots. The representatives of the biggest corporations staked themselves out nearest the doors of the Senate and House Chambers. No one knew if there was some protocol that gave the men with the biggest clients the best locations or if it was experience that gave the big hats the edge.

Introduction to the Legislative Process

Prepublication copies of this book were mailed to those who were thought to have an interest in the subject matter. Some who responded to a questionnaire expressed the opinion that a short primer regarding legislative procedures should be included for those unfamiliar with the movement of bills through the General Assembly. Following these suggestions, I have included this short description of the process.

An idea is formed by a member of the assembly or by someone who conveys an idea to an assemblyman. If the legislator believes in the merit of the idea, the legislator will make an appointment with a member of the Legislative Counsel's Office where a draft of a bill or resolution is made.

Once the final document is ready, the legislator will probably seek out others to cosponsor the measure before dropping the legislation in the hopper located at the front of the House Chamber, if the matter is introduced in the House. The same day the clerk of the House or an assistant clerk will assign a number to the measure and read the caption aloud in the House Chamber. The following legislative day there will be a second reader and the measure will be assigned by the Speaker to one of the standing committees.

In committee, one of the following may happen. The committee will decide the bill should not go forward and will give it a "do not pass" recommendation. Should this happen, almost without exception, the measure will not be called for a vote by the full House. The measure may be given a "do pass" recommendation and go immediately to the general calendar to be called or not called for a vote by the full House by the Speaker. It may be determined by the chairman of the committee in which the measure resides that a public hearing should be held. If the bill is controversial, it could be assigned to a subcommittee for further study. If the subcommittee does not report a "do pass" recommendation to the full committee or fails to report at all, the bill will not have a chance of passage.

The subcommittee might suggest changes that will enhance the chances of passage and the full committee might then pass the bill out with amendments as a committee substitute. But usually, referral to a subcommittee is the death knell for a bill.

In Georgia, according to the rules of procedure, a bill which has not passed by the last twenty days of a session, but has been given a "do pass" recommendation by a committee, must be presented to the Rules Committee for its action. Only if Rules recommends that the bill be placed on the general calendar is there a chance of passage. If Rules recommends "do pass," the bill goes to the daily Rules calendar and is left to the discretion of the Speaker as to whether it will be called or not. If not called, the bill goes back to Rules again.

Finally, if a bill is defeated by the membership of the full House, a motion for reconsideration may be made. The motion must, under the rules, be made at the proper time or not allowed. If the motion loses, the bill is dead. If the motion passes, the bill is returned to the Rules Committee, if the period of Rules is in effect or to the general calendar as the case may be. Only one motion for reconsideration is allowed.

When a bill is before the House for a vote, any member may offer amendments that are germane, and the full body will determine whether to agree to the amendment or not.

Immediately after a bill passes in the House, it is forwarded to the Senate. The same procedure begins again there. The bill is read two times and placed in a standing committee by the lieutenant governor, who presides over the Senate. Once in committee the bill may be given a "do pass" or a "do not pass" recommendation, or it may be placed in a subcommittee for further action or amended to become a committee substitute.

If the measure is passed by the Senate with amendments or by committee substitute it is then returned to the House. Back in the House, the following usually occurs. A conference committee is appointed, composed of three members of the House chosen by the Speaker and three members of the Senate chosen by the lieutenant governor. Those named to the conference committee must have voted in favor of the legislation in their respective bodies. If a bill is amended by the conference committee, it must be voted on by both houses. If a bill is not amended by the conference committee, it must be sent to the chamber in which it originated. When House Bill 212 (Coastal Marshlands Protection Act) was before the General Assembly, no conference committee was needed.

However, a conference committee can be avoided if, on a motion to agree with Senate amendments, a constitutional majority of the House is won. A constitutional majority is half of the membership of the full House plus one. If the motion is carried by a constitutional majority, the bill has passed and is forwarded to the governor for his approval or veto.

Office of Legislative Counsel

I made my way through the opening-day crowd to the Office of Legislative Counsel in an upbeat mood, expecting the final draft of my legislation to be ready. I hoped to put the Marshland Protection Bill in the hopper on the first day of the session, but now found that the draft was still in its earliest stages. Virlyn Slaton, the counsel assigned to do the drafting, was apologetic, stating that he had been assigned to an interim committee that had taken all of his time during the last few weeks.

I asked when he thought the bill would be ready and was given an estimate of a week. There was no way of knowing what direful consequences this and further delays might cause in the future.

We went over the draft of the first two pages. Few changes were ever made in the introductory wherefore clauses.

I stated to Mr. Slayton my firm belief that the state of Georgia owned a large majority of Georgia's marsh. I explained that in English history the king owned the tidelands and that a trust was impressed on the area for the use of the public. Any lands granted by the king remained subject to the trust unless specifically negated. Such negation was seldom, if ever, granted.

When the American Revolution ended, the lands that belonged to the English king became the lands of each of the original thirteen colonies and the same trust doctrine was imposed on them. I explained that I had read a number of land grants purporting to convey marsh titles. Most often the descriptions were so vague that the area attempted to be conveyed could not be found.

Finally, I told Mr. Slayton that I had been the closing attorney on many parcels of land in the coastal area, some abutting on the marsh. The highland was meticulously surveyed and the purchase price generally arrived at using a per-acre figure. The highland was warranted. The marsh was thrown in without warranty, without survey, without additional payment, and with only a cursory description. Returns on marsh for ad valorem state and county tax purposes, by the purported owners of the marsh, were generally at the rate of one dollar an acre.

I had provided counsel with the names of thirteen marsh grasses to be protected, having secured the names from Dr. Fred Marland of the Marine

Institute on Sapelo Island. I had also provided counsel with a definition of the estuarine area, which was defined as all tidally influenced waters, marshes, and wetlands lying within a tide elevation range from 5.6 feet above mean tide level and below. The elevation of 5.6 feet was not arrived at easily. It was the result of research by Marland, who with his wife, Sarita, used tide elevation benchmarks along the eastern seaboard as they drove with surveying equipment from Florida to Maine. The 5.6 feet above mean tide level separated the marsh from the upland.

I told counsel that I thought that the political subdivision to which an application to fill, drain, or dredge or otherwise alter a marshland had been submitted should consider, when deciding whether an application should be granted or not, whether the proposed activity would be harmful to a neighboring property owner, violate zoning ordinances, obstruct the natural flow of navigable water, or interfere with the propagation of shrimp, fish, oysters, clams, or other marine organisms or wildlife.

We last concerned ourselves with the question of penalties for violation of the act. It was determined that the first violation should be a misdemeanor, a fine not to exceed a thousand dollars or six months in prison. The second violation would be not less than a thousand dollars an acre.

I had previously, during our earlier correspondence, provided counsel with a list of those I wanted included on the state board which would oversee the provisions of the act. The proposed protection board would include representatives of the State Game and Fish Commission, director of the Water Quality Control Board, the Georgia Council for Preservation of Natural Areas, the Georgia Surface Mining Land Use Board, and the Health Department's director of Water Supply Service. These entities were chosen because of two factors—first because each one was environmentally concerned, and secondly because of my belief that during the course of passage, amendments might be offered which could weaken the board's composition, seeking to place members on the board that would not be friendly to marsh preservation. (Many suggestions came from numerous sources. Some wanted the Georgia Ports Authority included; others said to add the Georgia State Chamber of Commerce to represent industry. There were those who suggested that five members from the assembly, whose districts included the coast, ought to be named.)

I had also previously discussed with legislative counsel the question of what state department should be the standard-bearer or sponsor of the board. I had suggested the State Game and Fish Commission for several reasons, among which was the fact that the head of the commission as well as a number of his subordinates had been in favor of the legislation from the outset. Of equal importance was the difficulty one would face in creating a new free-standing entity that would require a separate item for budgetary purposes.

My hope that the bill would be ready for introduction on the first day of the legislative session on January 12, 1969, was dashed. Now I was required to spend the first two weeks of the session working with counsel toward completion of the legislation. I could see the Rules Committee looming in the distance.

One significant occurrence did happen on the session's first day. One of the leaders of the Georgia Federation of Women's Clubs, having heard of my interest in the marsh, invited me to a meeting of the club's executive council. I was delighted to go and speak to them of my plans for protective legislation. My talk proved to be worthwhile, since reaction was favorable and the organization provided great support as the bill moved through the legislative process.

It was on the same day that I met Jane Hurt Yarn. I soon discovered that we were related. My paternal grandfather was William Hurt Harris. During the next year and a half, Jane Yarn was a tireless advocate. She was forever on the telephone or sending letters and wires soliciting help from all parts of the state.

Her tenacity should be appreciated by every person who benefits from the many treasures the marshlands produce. It would be well, if during the course of reading what is written here, if one could remember Jane's name and the influence she had in all of the debates in the House and Senate and in the governor's final conclusion. If it were not disconcerting to the reader, every few pages I would from time to time simply add her name between paragraphs—Jane Yarn—just as a reminder that she was continuously at work behind the scenes.

On January 22, I introduced House Resolution 75-184 regarding marshlands ownership. The resolution authorized and directed the attorney general to conduct a pilot title analysis to determine the legal owners of marshland tracts in each of the coastal counties of Georgia.

The resolution was widely reported in the press, and I was sure that the attorney general was informed of its existence. Although the measure was finally voted out by the Natural Resources Committee, to which it had been referred, I never pushed for its adoption by the House. For one thing, the marshlands protection measure, together with other legislative matters, was consuming most of my time, and for another I feared that the resolution might blunt my strong stance that the state owned the marsh.

It was not until January 23, 1969, that the work was finished on House Bill 212. I put it in the hopper. Ten crucial days of the session had elapsed without the possibility of moving the legislation along. Thirty days of the session were left. The bill was read a second time on January 24, 1969, and the Speaker committed it to the State Institutions and Property Committee. I immediately sent copies to the Brunswick City Commission, the Glynn County Commissioners of Roads and Revenues, the Brunswick–Glynn County Chamber of Commerce, and the heads of numerous industries.

The response was immediate. None of those I contacted were in favor of the legislation. All were adamantly opposed and expressed their opposition by contacting the chairman and members of the State Institutions and Property Committee.

State Institutions and Property Committee

Because of the hue and cry, the chairman of the State Institutions and Property Committee stated that a public hearing should be held. He set the date for February 11, 1969, providing enough notice so that those who wished to appear would have adequate time to prepare. From January 24, when the bill was assigned to committee, until February 11, no action could be taken on the bill. In accordance with the Georgia Constitution of 1945, the General Assembly was required to take a break for a period of two weeks to allow the Appropriations Committee to complete work on the budget. In 1969, that break took place from January 27 through February 9. The General Assembly meets each year for forty days, Saturdays and Sundays excluded since the assembly is in recess under normal circumstances each weekend. The two-week break for the Appropriations Committee hearings is not deducted from the forty days allotted for the assembly's meeting time.

During this time, letters were being written and wires and resolutions were arriving at my desk in the House Chamber. Some of the correspondence was in complete opposition to the bill. The Brunswick–Glynn County Real Estate Board, the Brunswick Port Authority, the City of Brunswick, and the commissioners of Roads and Revenue of Glynn County expressed their views that the bill should be withdrawn. Some expressed the proposition that adequate laws already existed.

Prior to the hearing of the State Institutions and Property Committee in the House Chamber on the eleventh of February, 1969, the head of a large industry in my home community of Glynn County expressed strong opposition. The manager of Hercules Powder Company wrote the following letter. It was delivered to the president of the Brunswick–Glynn County Chamber of Commerce.

February 10, 1969
Mr. Frank Hill, President, Brunswick–Glynn County Chamber of Commerce, Brunswick, Georgia

Dear Frank:
Since I will be unable to attend the public hearing on the Coastal Wetlands Bill tomorrow, I would like to pass along to you the affect we anticipate if this bill becomes law. If this information will help you establish the Chamber's position at the hearing, you are welcome to use it as you see fit. Hercules Brunswick plant has been in operation in its present location for about sixty years. As the City has expanded, residential and commercial areas have gradually surrounded our parameters—[sic]. Recognizing the inevitable barrier to further expansion on the mainland, Hercules has obtained during the last twenty years or so title and options on marshland lying to the east of our present plant and contiguous with it.

The most recent acquisition of marshland occurred last year when in cooperation with the City of Brunswick and its Urban Renewal Program, Hercules agreed to give to the Urban Renewal plan property lying west of its parameter [sic] in exchange for marshland adjoining our eastern boundaries. This exchange was made in good faith in anticipation of future industrial use and would never have been consummated had we anticipated restrictions on its use as the Wetlands Bill would impose.

The economy of this operation, which currently employs over a thousand people in the area, is based on importing pine stumps from four southern states.

Adhering to these stumps is a small amount of dirt, sand or clay which cannot be removed prior to shipping. Last year this dirt amounted to over 15,000 tons. Up until the present, individuals and other agencies in this area have been using this dirt for fill and have been happy to haul it away in exchange for the loading by Hercules. Brunswick and the surrounding area today probably are a few inches higher because of this. Now that the area has been developed and roads paved, there is decreasing demand for dirt, and starting last year, we found it necessary for the first time to dispose of some of it ourselves. We anticipate that in the future, we will have to handle more and more of this. The only method available to us is to fill our marsh property.

If this means is denied us by this bill, the economy of this operation would be seriously and adversely affected and would cause us to look elsewhere when expansion is contemplated.

The stump wood which constitutes our raw material is currently procured in over fifty counties in southeastern and southern Georgia, and its purchase contributes over 2½ million dollars in the economy of these areas. The adoption of this measure would not be in the best interest of any of these counties as it would inhibit future growth of our harvesting activities. If I can be of further service in your efforts to prevent passage of this measure, please let me know.

Very Truly Yours,
Stan Fenelon, Manager

(A puzzling question remains regarding the swap between Hercules, Inc., and the city of Brunswick. Who, if anyone, certified title to the marshlands which the city deeded to Hercules in exchange for Hercules's highland?)

First National Bank of Brunswick

A copy of the Fenelon letter was mailed to my father and reached him after the committee hearing. The Fenelon letter certainly expressed a reason for concern. On the same day the Fenelon letter was written, my father, not having seen it, wrote to me.

February 10, 1969
A.M. Harris

Dear Reid,
The Directors of the Chamber of Commerce are going to meet Friday at noon and I wondered if it wouldn't be a good idea for you to attend the meeting and talk about your wetlands bill. There is so much conversation going on about the thing—probably largely by people who don't know what it's all about. I'm sure

you could put people's minds at rest, and as indicated, wondered if it wouldn't be a good thing to do.

With much love,
Dad

One would have to have known the man who wrote this letter for a very long time to understand how mild, how understated this letter was. It was written by a man who had been a banker from early in life, having risen through the ranks in banking from the lowest position to becoming the chairman of the board. He was immersed in the business community of a medium-sized town. His friends were leaders of the community, heads of industry. When he wrote, he served on the Georgia Ports Authority and was largely responsible for securing Colonel's Island for industrial development by the authority. One of his dreams was for the authority to establish a railroad to Colonel's Island and the tracks would have to transverse marsh.

I could feel his consternation. Here his son was plowing a path in opposition to the community's desires. Each day that my father sat at his desk at the bank, he was approached by a number of people expressing, in no uncertain terms, displeasure with his son. How difficult it must have been for him.

Earlier, during my legislative service, the legislature was considering a lease from the state to certain railroad rights-of-ways owned by the state. There were two bidders for the lease—the L&N Railroad and Southern Railroad. It appeared that Southern would win until it came to the attention of some legislators that Southern planned to drop service to certain towns along the railroad's route. When the top brass of the L&N made a commitment to maintain service to the areas Southern would drop, I joined with some of my colleagues to keep L&N as the lessee.

During the course of these proceedings, Mr. Brogdon, the president of Southern, called my father to complain of my actions. He told my father, the president of the bank, to get his son off of Southern's back or Southern would remove all of its funds from the bank. My father said that his son was his own man and that he would take no steps to deter him. Shortly thereafter the L&N won the lease. The president of Southern made good on his threat and removed all of the railroad's funds from the bank.

These are good illustrations of what family members sometimes bear for the public service of their kinfolks.

Following my father's request of February 10, I made plans to attend the meeting of the Chamber of Commerce.

Hearing in the State Institutions and Property Committee

In the meantime, in Atlanta, the public hearing was held at 2:00 p.m. on February 11, 1969, in the House Chamber at the Capitol.

February 11, 1969, was a difficult day for me. During the public hearing on House Bill 212, the House Chamber was fairly crowded. Members of the State Institutions and Property Committee were scattered among many interested observers, some of whom planned to speak.

Many were from the coast, particularly from my home district. The mayor of the city of Brunswick, the chairman of the Glynn County Commission, labor union leaders, heads of local industrial plants, and the Chamber of Commerce were present. Dr. Eugene Odum and members of the Marine Institute on Sapelo Island were there, including Dr. Fred Marland, Dr. John Hoyt, and Dr. Dirk Frankenberg. Representatives of government agencies had come: Rock Howard of the State Water Quality Control Board and Secretary of State Ben Fortson.

At the outset of the hearing, all who wished to speak were asked to sign a paper which had two columns, one column for those against the bill, and one column for those for it. The opponents represented industry, labor unions, and local government bodies.

Those favoring, in the main, were from the scientific community with knowledge of the value of the marshlands but also included those interested in esthetics such as the authoress Eugenia Price and members of women's garden clubs, including Mary Helen Ray of Savannah, Marie Melinger of Tiger, Georgia, and a club member from Banks County.

I sat at the front of the chamber near the rostrum from which each speaker made his or her presentation. I wished to see each one up close and hear what each had to say. I heard the manager of Brunswick Pulp and Paper Company decry the fact that if this bill were passed, the company's plans to plant pine trees in the marsh would have to be aborted.

I heard the Hercules Powder Company's plans to fill hundreds of acres of marsh with the sand and soil flushed from pine stumps used in manufacturing many of its products. From local government I heard that the bill flew in the face of home-rule provisions of Georgia law and that the local commissioners could handle all that was necessary in the way of zoning.

For the first time, from those favoring the bill, I learned that substantial dredging could cause the loss of significant amounts of oxygen in the water, thus endangering all forms of sea life in large areas in and far beyond the area of activity.

Dr. Odum spoke about the scalawags and carpetbaggers who owned the majority of coastal manufacturing plants and who were not concerned with the local environment.

Finally, the chairman was speaking, winding down the hearing, offering appreciation to those who had come.

"This hearing is adjourned," he said finally, dropping the gavel lightly on its block.

The crowd started to stir.

I stayed in my seat at the front of the House feeling utterly dejected. How much damage had been wrought by all the negative testimony? Could the bill be salvaged in the face of all that had just transpired?

Then a loud piercing voice reverberated through the chamber from the back. "Mr. Chairman," the voice called out.

"Why, yes, Mr. Secretary," the chairman said into his microphone.

"May I speak briefly to this gathering?" Ben Fortson, the secretary of state, inquired.

I was suddenly heartened. The secretary was a man of great influence. I knew he favored the legislation. I had talked to him in his office about it. He had shown me a plaque on his office wall that read: "It is better to have tried and lost than never to have tried at all." Mr. Fortson was the man who years before had had a special compartment built below the seat of his wheelchair in which to transport the great seal of the state of Georgia. The seal would not be used again until the debacle concerning who had been elected governor had been decided by the courts.

Mr. Fortson's wheelchair sped down the long, sloped center aisle as he exclaimed, "Sorry I arrived a little late. I didn't know about the roster."

"Certainly you may speak, Mr. Secretary," the chairman answered, and gaveled the audience to order. "Ladies and Gentlemen, I present to you, Georgia's esteemed secretary of state."

The audience settled down and the secretary turned to face them.

"I won't need the microphone, Mr. Chairman," the secretary said over his shoulder. "Spoken to larger crowds than this and was heard," he said.

Then the secretary turned back to the audience.

"For the record, let me make one thing abundantly clear. I wholeheartedly endorse Representative Harris's House Bill 212."

The secretary paused to scattered applause.

"I wouldn't change one dot on an *i* or cross on a *t*. I am also, by the way, in absolute accord with Representative Harris's resolution to determine the extent of our state's marsh ownership, but that's not the question before us today. As your secretary of state, I say to you that we must stop the encroachment on our marsh, and I must tell you of a dereliction of my duty to you. Few people know that among the titles I hold by virtue of my office is that of surveyor general. In that capacity I have been woefully negligent.

"Over the years I should have made it my business, because it's the people's business, to determine the extent of the state's ownership of the coast. I haven't done so, haven't requested the funds to do so. Now, because of my failures, Representative Harris here has taken on a tough job. Mr. Chairman, this bill was placed in the right committee. You are dealing with state property here on this. I implore you and the other committee members to take quick action in giving this bill a 'do pass' recommendation. I solicit your aid in supporting it through the House. Should this measure fail, the coast will quickly suffer dire consequences. Many who reside there now will have to move out because of the devastation that will come if some meaningful protection, such as that offered by the bill, isn't immediately forthcoming. I fervently pray you'll heed my plea and support this measure."

Because of his reputation and the esteem the members of the General Assembly had for him, the secretary's testimony was most helpful toward the passage of the marsh legislation.

I casually mentioned that the author Eugenia Price was among those testifying at the hearing of the State Institutions and Property Committee in the House Chamber. Later I will set out the nature of her testimony when I describe her appearance before the Senate Industry and Labor Committee. She brought to the House committee a petition signed by two hundred or more residents of St. Simons Island favoring the passage of the marshlands legislation. The petition had been prepared by Miss Price and her companion Joyce Blackburn, and the two of them had taken considerable time in securing signatures in various places on the island.

After her testimony was concluded, Miss Price went later in the day to the Atlanta airport for her return flight to St. Simons Island. Standing in line, waiting to board the plane with her, were some coastal businessmen and local politicians who were fighting the legislation. All of the men knew her. They gave her a unanimous cold shoulder. Their coldness impressed her so deeply that she wrote of the experience in her book *St. Simons Memoir* nine years later.

The same attitude of disdain, mainly from the same sources, was to follow me for a substantial amount of time. Only with the passage of time, when the need for estuary protection became better understood, did a change occur. Gradually I was stricken from the list of those who were deemed to be against industry, against labor or progress.

The only means I had of coming to grips with the evident displeasure of those who were opposing the marshlands legislation was the firm belief that, eventually, those in opposition would understand the inestimable value of the area I was attempting to protect. I had the opportunity of being instructed concerning the value. Most of the public had not had the opportunity for such instruction.

At the end of the hearing, Phillip Chandler, the chairman of the committee, determined that the bill was so controversial that it should be placed in a subcommittee for further study. This was a bitter blow. A subcommittee was often a bill's burial ground.

On February 13, 1969, my father wrote the following letter:

Mr. Stanley Fenelon, Manager
Hercules, Inc.
Brunswick, GA 31520

Dear Stan:
I appreciate your having sent me a copy of your February 10 letter addressed to Frank Hill, President of the Chamber of Commerce, about Reid's marshland bill. Myddelton, Jr. and I have talked to Reid quite a lot about the bill and find he is dedicated to what he is doing and is definitely of the opinion that no local industry, or a potential one, would be penalized by its passage.

What I understand is that he hopes to provide some regulation of the broad areas of marshes adjacent to their plants. He expects, I am sure, that this will be covered in the rules laid down by the board appointed. I am sure you know that, in case of an arbitrary ruling by the municipality or the board, an appeal to the local Superior Court is provided for.

As you know, I have been serving as a member of the Georgia Ports Authority, to locate an industry on Colonel's Island, building a railroad to it. We are now organizing with a group of Texans to build a refinery there and the plans contemplate erecting a plant on 132 acres of marsh. I am not at all concerned about our ability to have the marsh used for that purpose.

As indicated, I am sure Reid does not intend that any harm come to your good company and I believe you could be reassured by talking to him.

Sincerely,
A. M. Harris

(Copies of this letter were sent to G. E. Bosserdet, Hercules, Inc., Wilmington, Delaware; Mr. Frank Hill, Brunswick–Glynn County Chamber of Commerce; and to me.)

At the time of the hearing by the State Institutions and Property Committee, there was a steady flow of favorable comments that came from all over the state of Georgia. Words of encouragement came from Atlanta, Columbus, Savannah, Rome, and many other places, including my home district of Glynn County. Everyone who was favorable to the legislation was of the opinion that the marsh was a highly valuable resource.

My overwhelming problem was to find a way for protection that would be palatable to most, yet have enough strength to be meaningful.

State Institutions and Property Subcommittee

After the hearing on February 11, 1969, and the notification by the chairman of the State Institutions and Property Committee that he had placed my bill in the hands of a subcommittee, I began to talk to a number of members of the House to get their reaction to the bill. Almost without exception I found that the main concern was that the legislation was tampering with "private property rights."

For the next few days I pondered what could be done to overcome these concerns. These concerns were greatly emphasized when, at my father's suggestion, I met with the members of the Chamber of Commerce in Brunswick. Those at the meeting were not hostile, but there was obvious displeasure with my legislative agenda.

Then, one evening back in Atlanta in my apartment, sitting at the dining-room table with a legal pad in front of me, I began to write a committee substitute for the bill.

I struck out the provisions for a marshland agency and provided instead that anyone who wished to disturb the marsh by filling, dredging, or altering it, should petition the local government authority for a determination that the application was not in violation of any local ordinances or contrary to the general health and welfare of the community in which the marsh was located.

Once that determination was made, I provided that the applicant would be required to submit a title insurance policy, showing that the applicant owned the marsh to be affected, to the secretary of state of Georgia who would issue a permit to proceed with the applicant's plan. In writing this substitute, I wrote with unusual speed, feeling all the while I was somehow being led by a source outside myself.

I had been getting some rumblings from the Georgia Highway Department regarding the bill. At the same time lobbyists from the Georgia Power Company and Southern Bell were expressing dissatisfaction. Building of roads and construction and repair of utilities were vital to everyday life. The Highway Department and the utilities acting in concert would be powerful forces against the passage of the bill.

I determined that exemptions should be provided. I noted on my legal pad a list of exemptions and numbered them: (1) State Highway Department, (2) United States and state agencies responsible for navigation, (3) public utilities regulated by Public Service Commission, (4) railroads, (5) water and sewer by political subdivisions, (6) private docks on pilings above marsh by owners of adjoining highland.

I felt that these exemptions would not appreciatively weaken the act.

More time was to pass while I had copies of the proposed committee substitute retyped by the Office of Legislative Counsel. I don't know the date that the substitute bill was ready to present to the State Institutions and Property Subcommittee. I only know that the subcommittee members approved my new proposal unanimously on February 18. I still retain a copy of the substitute signed by the members of the subcommittee, the signatories being Reps. Al Burruss, Herb Jones, P. Evans, and myself.

On Friday, February 21, copies of the proposed substitute were distributed to members of the State Institutions and Property Committee.

On February 24, at a meeting of the full committee, Chairman Chandler picked up the original House Bill 212 and turned it back and forth examining its cover.

Turning toward me, the chairman asked, "Have you seen this lately?"

"Seen what, Mr. Chairman?" I answered.

"Your bill. Take a look. You're losing supporters."

The Chairman handed the bill to me. I was stunned to see that the names of several House members who had signed as cosponsors had been scratched out.

"When did this happen?" I asked.

"Don't know. When I picked the bill up from the Clerk's Office today, I noticed that names had been lined through," he said.

As I sat there in a state of shock, a member of the committee asked to see the bill.

"Several names have been stricken. Some of them are from the coast. Mr. Chairman, do a majority of the members from the coast want this thing or is there a majority against it?" he asked.

"I haven't made a count," the chairman said.

"Well, seems to me this is a bill that affects only the coast. Shouldn't it be treated like local legislation? Won't the coastal delegation have to be unanimous on this? That's the rule in the Local Affairs Committee if you're going to get a 'do pass' recommendation. Everybody representing the area affected has to be in agreement," the committee member said.

I started to voice my objection to this line of thinking but was cut off by the chairman.

"I hadn't planned to discuss your bill today—only wanted to show you the desertion from your ranks. We'll take up discussions on House Bill 212 tomorrow, when, hopefully, more committee members will be present," he said.

As the meeting drew to a close, I announced that I had copies of the proposed substitute for anyone who had failed to get a copy on Friday.

The chairman was gathering his papers and preparing to leave when I saw Herb Jones, a member of the Chatham County delegation and of the subcommittee approach the chairman's table.

"Let me see the original House Bill 212," he said. And taking out his pen, he signed the bill with a flourish. "I didn't get to sign it before. Better late than never. If some of my colleagues from Chatham area are going to desert, I'll take up the slack," Representative Jones said.

Representative Jones's actions temporarily raised my spirits, but for the balance of the day, during the evening and the following morning, I was concerned over the specter of needing either a majority or the unanimous approval of the coastal representatives on the committee to get a "do pass" recommendation. There was not a prayer for that.

The next day, February 25, my worries proved to be groundless. No one mentioned majorities or unanimity. The first order of business was House Bill 212. The chairman announced the unanimous approval of the subcommittee for the proposed substitute and called for a vote. A majority voted in favor of passing the bill out of committee. Now the bill would go on the general calendar and the Speaker must be convinced to call it up for a vote of the full House. Time was of the essence, since the Rules Committee would become involved on March 7. I had ten days.

House Vote on House Bill 212

Each day from February 25 to March 5, with the exception of the weekend, I pleaded with the Speaker to call House Bill 212 for a vote, without success. On March 5, the Speaker finally indicated that this was the day. The calendar was lengthy, and the wait seemed interminable as other measures were debated into the afternoon. As time passed, several members of the House approached me to say they were for the bill but had other engagements and could not wait longer to vote. Arthur Funk from Savannah, a staunch supporter, was one of the last to leave before the Speaker called action on the bill.

That afternoon turned into a nightmare. Some students from Georgia State University had called a rally to protest the war in Vietnam on the grounds of the Capitol. A number of House members had left their seats to look out the windows or to go outside to see what was happening. As I walked down the aisle to the well of the House, I saw many empty seats. As I approached the front of the chamber, Elliott Levitas, a cosponsor of the bill stopped me.

"The House is too light," he said. "Tell the Speaker to wait until tomorrow."

"I can't," I said. "I've got to get this to the Senate."

I don't remember much of what I said in the well that day. I recall my first words were: "As God is my witness, I wish I weren't standing here today."

I explained that members of my home community, including the mayor and council of the city of Brunswick, the Glynn County Commission, the local Chamber of Commerce, and the head of local industry, had objected to the bill but that I thought it much too important a matter to give it up. I expressed my belief that personal political gain was secondary to fighting for what I felt strongly to be in the public's best interest.

I'm sure I explained the value of the marshlands, stressing the uniqueness of the Georgia marshes, their productive capacity, and a production equaling ten tons of organic material per acre each year in comparison with seven and a half tons produced annually by one acre of corn in the United States.

I talked about the nursery grounds for shrimp, crabs, oysters, clams, and fish. I described the marsh as a pollution fighter and a buffer against storms. I told of the sad plight of the marshes of New Jersey, Massachusetts, Maine, New Hampshire, Connecticut, Rhode Island, and San Francisco Bay that had been lost forever to filling and dredging.

I told the members of the House of my firm belief that the Georgia marshes belonged to the people of the state and for the people should be safeguarded. I must have said something about the uniqueness of the marsh growing area, arising from the proper temperatures of air and sea, the protection afforded by the out-jutting Florida peninsula and the northward and eastward curve of the Georgia coastline.

Rep. Mike Egan of Atlanta made a short seconding speech seconds before the bill came to vote.

"A person should show that he owns property before he goes in messing it up," Egan said.

Now, almost unbelievably, the Speaker's voice boomed through the sound system.

"The question now is, shall House Bill 212 by committee substitute now pass? The clerk will unlock the machine."

I was walking up the slanted aisle toward my seat at the rear of the House Chamber when the voice of Robert Harrison, the representative from Camden County, called out from my left.

"Mr. Speaker, I have an amendment," Harrison said.

As I reached my seat the Speaker turned to the clerk.

"The clerk will read the amendment," the Speaker intoned.

Not knowing what might be coming, I feared the worst.

The clerk read the amendment.

"The next to last section shall read: 'This Act shall become effective July 1, 1970.'"

"Is there objection?" the Speaker asked.

I was stunned. Should I object? My mind went into high gear. What Pandora's Box would I be opening by an objection?

Looking at me, the Speaker asked again, "Is there objection?"

"No objection," I said.

"Shall this bill now pass by committee substitute, as amended? The clerk will again unlock the machine." The Speaker dropped his gavel on the rostrum and sat in his high-backed chair.

The ayes were ninety-six and the nays were fifty-nine. Needing a constitutional majority of ninety-eight votes, the bill lost by two votes.

I stated that at the proper time the next day, I would make a motion for reconsideration of the House's action in failing to give House Bill 212 a constitutional majority.

At the end of the legislative day, I sat in my seat watching my fellow legislators putting on their coats and gathering their belongings to the leave the House. There was an awful sinking feeling arising from the narrowness of my loss. I admonished myself for not taking Representative Levitas's advice to postpone the presentation of the bill.

While sitting there, a striking young female came up and sat in the seat next to mine. She had an alluring body. The faint odor of expensive perfume drifted with her as she passed behind my desk.

"Bad day," she said.

"Yes," I answered. "It was a very bad day."

"Are you going to try again?"

"You bet I am."

"This may sound crazy and a little forward, but what are you doing tonight?"

"Why do you ask?"

"I was thinking you might come to a little party after a day like this." A beguiling smile was on her lips: a dark-haired, mischievous temptress flouting her wares.

"I'm sorry. My wife and one of our sons are coming to town." We were planning for them to come the following week, so my answer was truthful.

"How about coming tomorrow night?"

"I'll be heading to the coast then."

She got out of the chair and strutted away without another word.

I saw her only one other time. She sashayed down the center aisle, bent over and whispered something in a representative's ear seated near the front, then quickly turned and left. Several minutes later the representative followed her from the chamber. It was early in the afternoon when she arrived. I noticed that the representative never returned to his seat that day.

I have often wondered who had hired her to spread her mischief.

After my visitor left, I went to the Clerk's Office and got a copy of the tally sheet made by the voting machine that shows the number of aye votes, the number of those not voting, and the number of nay votes. The sheet also shows whether the bill originated in the House or Senate and the number of the bill or resolution. On the lower right corner, the date is shown. The names of all members of the House are set out in alphabetical order. Black dots indicate how each member voted. Dots on the left of a name indicate an aye vote. A dot on the member's name indicates no vote was cast and dots on the right of the name indicate a nay.

By studying each member's vote, I could to some degree determine which members to approach, try to convince some nonvoters to vote for the bill, and some who voted nay to change their minds. Of course there were some with whom I would not waste my time.

I don't know how many votes changed with my continuing explanations, but I believe there were a few.

One day, while going from one member of the House to another in the attempt to secure votes, I was told that the seat of my pants had ripped open leaving a large gap. I went to my seat and thought about what my next move should be. After a while I moved down to the front of the House and sat in the seat next to a member of the House who had made an unusual argument during the debate on Georgia going on daylight saving time. His argument against was that if the measure passed, there would be an hour less sunlight for Georgia's crops, and Georgia's farmers would be in peril.

I had chosen this member with a particular purpose in mind. Of all the representatives he had been the most reserved toward me. Any time I spoke to him he scarcely acknowledged my presence. I had wanted to gain some relationship with him but had little success. Coaxing one to another's point of view is called politics and I needed every vote I could get.

I explained to the gentleman, who shall remain nameless, that the seat of my pants had ripped open and he was immediately sympathetic. "Hold still where you are, young fella," he said. "I'll get you some help."

"How?" I asked.

"You just stay sittin' there, brother. I've got a phone call to make."

Soon he returned. "Help's on the way," he said.

"What kind of help?" I asked.

"You'll see. You'll soon see."

A few minutes passed. Then a state trooper, dressed in full regalia, was standing by my side.

"Come with me," he said.

"Where are we going?" I asked.

"Going to wherever you live so you can get a good pair of britches."

Strangely, after that, my heavyset member of the House, who had the weather-beaten appearance of the farmer he was, came over to my camp. A rough-and-tumble man though he was, he was reelected many times to his post.

One never knows what will form a relationship.

On March 6, I moved reconsideration and the vote was 101 to 21 in favor. The reconsideration vote gave me one last chance. One motion for reconsideration is all that is allowed. House Bill 212 was still alive by committee substitute.

From March 6, 1969, when my motion for reconsideration was favorably voted, until March 14, I spent a part of each legislative day talking one-on-one with House members, checking off names of those with whom I had conferred and attempting to keep the score on the members I could rely on for support.

Now, I went to the Rules Committee with my prayer rug, asking that Rules place the bill on the calendar for a House vote.

In the meantime the newspapers around the state were reporting on the fact that the bill had lost. A short editorial in the *Atlanta Constitution* read:

STAND UP!!

A bill to provide safeguards for the uses of Georgia's Marshlands has failed in the House by two votes. The legislation will undoubtedly be reconsidered. Will two gentlemen stand up for beauty?

Sometime after the House failed to give the marshlands bill a constitutional majority, and at the time when I began appearing before the Rules Committee to get the measure moving again, my wife, Doris, and I were having breakfast at the Atlanta Athletic Club. As our breakfast was being served, Rep. Robert Harrison of Camden County came in and in passing paused at our table. I introduced him to Doris, we said a few words to each other, and he quickly went on his way with a group of friends to sit at a nearby table.

Always protective of me, Doris had given Harrison a withering stare, a frightful stare our boys know well but cannot accurately describe. She spoke not a word in acknowledgment of the introduction.

As Harrison walked away, her eyes bored into his back. "Why is he fighting you?" she asked. "I'd like to give him a good swift kick in the pants," she said.

Doris was frustrated. I was frustrated. Frustration comes often to one familiar with the rules of the House. Greater frustration arises when one's mate, unfamiliar with those rules, begins to prod to move things along. In the mate's opinion, his or her spouse is doing the right thing, the perfectly logical thing. If the subject is legislation, it should be passed and signed into law in short order without any folderol. The big question becomes, "Why is this taking so long?" When the matter becomes protracted, watch out. Arguments can arise and become bitter from time to time. Once in a while, this happened to us. This was another price paid for public service.

I changed the subject as quickly as I could. "Are you still planning to get an associate degree in nursing?" I asked.

Rules Committee

Friday, March 7, was the first day the Rules Committee met in 1969. The committee met in a room on the fourth floor of the Capitol building. The room was not large. It had a conference table, and there were chairs at the head and foot and on each side. Since there were a large number of members of the committee, there were not enough seats at the table. Other rows of chairs flanked the walls and were used to seat committee members arriving after the first arrivals had taken seats at the table.

The room was greatly overcrowded. Representatives and senators appearing before the committee had to stand by the door, and some stood outside the room waiting their turn to enter after others had made their presentations and left. During hearings, the room was always crowded and always hot. There were no windows to open for cool air and the heat from the radiators was always on full blast. The room reeked of cigarette smoke.

The atmosphere added to the nervousness of all who appeared with a request to have their bills placed on the calendar. Naturally, the members of the committee who had bills in which they were personally interested were heard from first.

The Rules Committee voted "do pass" for the committee substitute, but the Speaker did not call it for a vote. This meant I would have to appear before Rules again. Even though a bill is placed on the Rules calendar one day, it may not be approved by Rules to appear on the calendar another day.

A weekend intervened. On Monday morning, March 10, my colleague from Glynn County, Richard Scarlett, presented the chairman of Rules with a telegram he said he had received from the Brunswick City attorney, Robert Sapp, and the city manager, J. Edward Hulse. The wire read as follows:

> Urge that bill 212 and all amendments thereto be not passed for session that in our opinion the Bay Street project would be jeopardized resulting in a tremendous loss to the improvement of Brunswick and would impair 15 years of planning and development of Newcastle Street, (341) and its connecting link to Bay Street would not allow its true and best use. Law as proposed is not needed in Brunswick and Glynn County. Zoning laws clearly protect marsh.

George Busbee, the chairman of Rules and later governor of Georgia, asked if Richard Scarlett, the second representative from my county, had made me aware of the telegram.

"No," I said. "When did he give the wire to you?"

"Earlier today," Busbee said.

"I see. Strange—he's been seated in the House Chamber in the seat next to me most of the day. Maybe it slipped his mind," I said with a smile.

"You think so?" said Busbee.

"I doubt it," I replied.

The chairman wanted to know my response to the wire. I responded that there was no marsh on Bay Street and I didn't believe that Newcastle Street encroached on the marsh. I said that even if some marsh were involved, the state and federal governments would require good title before proceeding. I also explained that thousands of acres of highland were available nearby for industrial expansion and that the area already had a multimillion-dollar industry that depended on the marsh for the production of seafood.

The bill was given a "do pass" by the Rules Committee each legislative day from the tenth through the fourteenth of March. It had been necessary on each of these days to appear before the committee.

Second Vote by the House

Yet, each day through the thirteenth, the Speaker failed to call for a vote. Then, almost unbelievably, on the fourteenth, the Speaker's voice boomed through the sound system.

"House Bill 212 by committee substitute is now before us. The gentleman from Glynn is recognized."

Suddenly, I found myself in the well of the House again. I had considered carefully what I should say this time, this last time to convince the House to approve the bill. I reminded the House that two committees had given a "do pass" recommendation to the legislation.

Both the State Institutions and Property and Rules Committees had thought the bill had sufficient merit and had given their stamp of approval. I told the membership of the House again that there were objections to the legislation, the most vocal being from my home district of Glynn County. I named the city commission, the county commission, the Chamber of Commerce, local industries and labor unions, and the Brunswick-Glynn County Real Estate Board.

Why remind the House of opposition? Wasn't that a foolish thing to do? I didn't think so. First, because many members had already heard speeches in opposition and most others had received letters and wires from those I named as my opposition. Second, because it showed without a shadow of a doubt that I thought the legislation was of vital importance by my willingness to buck that opposition.

I went on to say that the House should consider the fact that those who opposed had not had the same opportunity of working with the scientists who had studied the marsh environment that I had over the last several months. The fact was that most of the community believed the marshlands to be worthless, an area to be filled and built upon. Until being informed to the contrary by those investigating the productive capacity, and by reading literature concerning the estuarine area, I had shared the common "worthless" notion.

I must have mentioned the value of the marsh landscape again, being careful not to talk too long. I reminded the membership of the role that the marsh played in seafood production and the industry it provided. In the end, I said that I believed with my whole heart and soul that the legislation before them was vital, needed legislation.

I can still hear the Speaker's announcement after the votes had been cast.

"On the passage of House Bill 212, the ayes are ninety-seven and the nays are thirty," he said.

I fully expected his next words to be: "This bill having failed to receive a requisite constitutional majority has therefore lost."

Instead the Speaker said, "The Speaker votes aye. This bill having received a constitutional majority has therefore passed. The gentleman from Glynn moves that this bill be immediately transmitted to the Senate. So ordered."

And with the bang of the gavel, the legislative day ended.

Speaker George L. Smith, the first speaker in modern times to be elected by a caucus of House members voting by secret ballot. His crucial vote on the Coastal Marshlands Protection Act provided the majority necessary for passage and transmittal to the Senate. Painting by George Mandus, Capitol Art Collection, 1972. Reprinted by permission of the Georgia Archives.

Jane Hurt Yarn was a pioneering conservationist whose most important early contribution was her successful organization of the state's garden club members to write thousands of letters to Governor Lester Maddox's office opposing the strip-mining of phosphate on Georgia's barrier islands and supporting the Coastal Marshlands Protection Act. Yarn is shown here at a press conference with Governor George Busbee in 1976. Photo courtesy of Dr. Charles P. Yarn.

Ben W. Fortson Jr. served as Georgia secretary of state from 1946 until his death in 1974. His testimony at key legislative hearings was strongly supportive of marshland legislation. His contacts with members of the Georgia General Assembly were vital in furtherance of the passage of House Bill 212. Courtesy of the Georgia Archives, Small Print Collection, spc18-023.

Eugenia Price, best-selling author of a trilogy of novels set on St. Simons Island (*The Beloved Invader*, *New Moon Rising*, and *The Lighthouse*), gathered petitions among her fellow coastal residents supporting House Bill 212 and testified before the State Institutions and Property Committee and the Senate Industry and Labor Committee. Photo courtesy of Robert Cotner.

GARDEN GATEWAYS
OFFICIAL BULLETIN OF GARDEN CLUB OF GEORGIA, INC.

SUMMER ISSUE

Long active in state and national garden clubs, Mary Ellen Ray of Savannah was one of several key citizens prominent in their communities who directed others to advocate for the passage of House Bill 212. Courtesy of the Hargrett Rare Books and Manuscript Library, University of Georgia.

Left to right: First Lady Hattie and Governor Lester Maddox at the Governor's Mansion with the author's wife, Doris, and the author. Doris was a powerful advocate for protecting the coastal marshes by securing petitions from local PTAs and convincing Eugenia Price to testify before legislative committees. Photo courtesy of Michael Harris.

University of Georgia biologist Eugene Odum, shown here in 1996, was an innovator in the field of ecosystem ecology. Named a member of the National Academy of Sciences, he founded the UGA Institute of Ecology (later named after him) and two research stations, the UGA Marine Institute and the Savannah River Ecology Lab. He appeared before House and Senate committees speaking on behalf of House Bill 212. Photo courtesy of Betty Jean Craig.

Senators of the Georgia Coastal Marshlands Committee huddle before the start of a meeting in the capitol in January 1970. Present were (from left) Senator Robert Small of Griffin; Senator A. W. Holloway of Albany, chairman of the committee; Senator John Riley of Savannah; the author; and Senator Bob Walling of Atlanta. Senator Armstrong Smith (far right) attended the meeting, although he was not on the committee. Reid Harris Papers related to the Coastal Marshlands Protection Act, Richard B. Russell Library for Political Research and Studies, University of Georgia Libraries.

Two public hearings held on the proposed lease by the State Mineral Leasing Commission in Atlanta and Savannah were jammed with opponents of phosphate mining. Coastal residents, commercial and sport fishermen, professional conservationists, and citizens were almost unanimous in their opposition to leasing state bottomlands. Photo by Ted Borg, Georgia Game and Fish Commission, 1968.

Governor Lester Maddox signed into law Georgia's first legislation to protect the state's coastal marshlands on March 27, 1970. Eugene Odum, Jane Yarn, Rock Howard, and the author were among the observers. Photo courtesy of Michael Harris.

Left to right: The author with his father, Augustus Myddelton Harris Sr., and his brother, Augustus Myddelton Harris Jr. As the chairman of the board and president of the First National Bank of Brunswick, respectively, they heard much criticism of their son and brother without ever attempting to dissuade him from his goals. Harris Sr. served on the Georgia Ports Authority in the 1960s and 1970s. Harris Jr. was president of the Georgia Bankers Association the year House Bill 212 passed. Photo courtesy of Michael Harris.

III.

The Bill in the Senate

I thought I had an agreement with the lieutenant governor, whose name was also George Smith, but with a middle initial of T, that House Bill 212 would be assigned to the Senate Judiciary Committee. Among the Judiciary Committee membership were several staunch supporters of the bill who wanted to march it through a Senate vote. I sought out the lieutenant governor and told him again that assignment to Judiciary would be very helpful. He said that he would do what he could; not a strong commitment this time, to my consternation.

On March 17, three days after passage by the House, the bill was referred by the lieutenant governor to committee. He switched and chose to put it in Industry and Labor. I had a terrible feeling that the bill would die there.

On March 21, 1969, I received a note from the chairman of the Senate Industry and Labor Committee, Al Holloway, stating that the next meeting of the committee would be held on Monday, March 24. I knew then the bill had no chance of passing in the 1969 session, which would end March 26.

The depth of my disappointment is indescribable. Now months of uncertainty loomed ahead with Senate action pending.

Spring and Summer 1969

Sometime after the legislative session in 1969, I looked over a number of newspaper clippings reporting on and editorializing about the progress of House Bill 212. I was reminded that the editors and reporters had stated over and again that the bill was much weaker than the bill had been in its original form and that the bill the House had passed was a "watered-down" version. I have earlier referred to the words *jus privatum* and *jus publicum*. I want to stress that these two phrases are very important when it comes to understanding rights where coastal waters are concerned. What is private? What is public? What does the trust doctrine provide with reference to the coastal zone?

A summary of the trust doctrine was described by the United States Supreme Court in *Shively v. Bowlby*, 152 U.S. 1 (1894), as follows:

> By the common law, both title and the domain of the sea and of rivers and the arms of the sea where the tide ebbs and flows, and of all lands below high tide mark, within jurisdiction of the Crown of England are in the King. Such waters, and the lands which they cover, either at all times, or at least when the tide is in, are incapable of ordinary and private occupation, cultivation and improvement, and their natural and primary uses are public in their nature, for highways of navigation and commerce, domestic and foreign, and for the purpose of fishing of all the King's subjects. Therefore the title, jus privatum, in such lands, as of waste and unoccupied lands belonging to the King as sovereign; and the dominion thereof, jus publicum, is vested in him as the representative of the nation and for public benefit.

Since the original states of the United States succeeded to the king's status as trustee, the doctrine above annunciated can best be understood by substituting the word *State* or the words *State of Georgia* for the word *King* as *King* was used in the above opinion.

I was familiar with the *Shively* case and others like it. I was familiar with the trust doctrine. If the state of Georgia, as owner of the marsh, held it in trust for the public use, how could any individual claim to own it? I thought none could. By rewriting House Bill 212 as a committee substitute which provided for the securing of a title insurance policy, the bill was not weakened but was substantially strengthened.

As mentioned before, the committee's substitute bill which passed the House requiring title insurance was referred by Lt. Gov. George T. Smith to the Senate Committee on Industry and Labor, chaired by Sen. Al Holloway of Albany, Georgia. Few if any of my fellow legislators, certainly less members of the general public, understood that no title insurance company would issue a policy insuring private ownership of the marsh. Without a title insurance policy, no alteration of the marsh could be legally done. From the time the committee substitute was presented to the State Institutions and Property Subcommittee and until the fall of 1969, no one ever raised the issue with me.

On May 14, Jane Yarn wrote a letter to me. It read in part:

> I'm real ashamed that I haven't written before, because certainly you have been on my mind enough. I think the courage you showed during the 1969

session, and your hard work to get results for what you believed in was a wonderful inspiration. I'm sorry the results were not better, but with the lack of understanding and pressures from other sources it is understandable. We must gather our forces and put up a better fight next year. I'm glad we're on the same team.

On Tuesday, May 27, 1969, an article headlined "Kerr-McGee 'Not Backing Off' in Phosphate Mining" appeared in the *Brunswick News*. The article read in part:

> Kerr-McGee Corp. officials said Monday that the company was "not backing off" from plans to mine phosphate on Little Tybee and Cabbage Islands in Chatham County, and would possibly reopen negotiations for other state-owned islands in the area.
> ....
> A. F. T. Seale, senior vice-president of Kerr McGee, said its firm is "keeping its mining plans alive" despite Savannah being "hostile to phosphate mining,"

This article spurred me once again to action.

Senate Interim Committee

During the summer months of 1969, I wrote several times to Sen. Al Holloway asking that a hearing be held on the coast on the marshlands bill. It was not until September that I finally got a response. The time was then set for October 17 and 18, 1969, for the Industry and Labor Subcommittee to meet. Scientists at the University of Georgia Marine Institute conducted a seminar on Sapelo Island for the subcommittee explaining the values of the marsh and then public hearings were to be held the following morning at Brunswick Junior College.

Prior to the meeting of the Senate's interim committee for the Sapelo briefing, one of the interim committee members, Sen. John Riley, who had recently received notice of the meeting, wrote to me, and to Reps. Mike Egan and Elliott Levitas, expressing his concern with the title insurance requirements of the bill. His letter was dated September 24. I answered with the following:

> Dear Senator Riley:
> I appreciate your letter regarding the title policy provisions in House Bill 212. There is no question that in many instances securing title insurance to marshlands in Georgia would be very difficult unless the applicant for it secured a

deed from the State. The main question is simply that the State of Georgia in many instances has a better claim to the title than the person claiming the land.

As a practicing lawyer I have closed many real estate transactions over the years here in the coastal area. It is of great significance to me that in all the years of my practice no request has been made for title examination to a marsh tract unrelated to purchase of high land. Also of great significance is the fact that purchasers and sellers have made their deals based on the amount of high land involved. There is no relationship of purchase price to the amount of marshland "thrown in" on the deal. Also of significance is the fact that deeds purporting to convey marsh, except in rare instances, warrant the high land and quit claim the marsh. Buyer and seller know that the title to the marsh is questionable.

Should it be possible for an individual or corporation to utilize lands not belonging to them? I think not.

On the other hand, should legitimate industrial, commercial, or residential developers be absolutely precluded from the utilization of marsh areas? I think not. Development of urban marsh tracts may be in the best interest of the people of the State. Where development is best, I am sure the State, through presently existing agencies under State law, (which provide for the method of sale of State owned lands) would grant or sell State owned marshlands to developers. This would be done by the same method established by law for the sale of all other State properties.

We would not only keep squatters from appropriating what belongs to all the citizens of the State, but would hopefully have a vehicle for orderly development of the area, since the State, in every case where title was doubtful because of State ownership, would decide whether the price paid to it for the marshland was sufficient return for the great productivity lost in terms of food supply, recreational and aesthetic uses.

Finally, there would be one central place, the Secretary of State's Office, for the availability of information which would tell us how much of the marsh we were losing each year to development. Now we have no idea how many acres are being lost to filling, dredging, draining, etc. I look forward to discussing this further with you on the 17th and 18th.

Reservations have been made for you, etc.

Yours Sincerely,

Reid W. Harris

Senator Riley missed the interim committee meeting due to illness of his wife. That my letter to him did not persuade him is evident from a letter he wrote on December 16, 1969. His letter read in part:

I have studied House Bill 212 quite thoroughly, and although I fully concur with its objective, it is my opinion that the prerequisite of obtaining title insurance before the Secretary of State can issue a permit makes the Bill completely unenforceable.

The Ocean Science Center of the Atlantic Commission is currently embarking on a project to attempt to come up with a bill to protect our marshlands and islands. Unfortunately, this study will take approximately a year; however, in my opinion, it is the proper approach to take to this serious problem.

As it turned out, not only Senator Riley but Sens. Bob Smalley and Bob Walling also were unable to attend the committee meeting on October 17 on Sapelo Island, which meant that two of my strongest supporters were missing. Once again I felt discouraged. Senators Walling and Smalley had agreed to introduce a bill in the Senate identical to the one I had introduced in the House.

Sapelo Island is a beautiful place, one of the gems of the Golden Islands of Georgia. Some ten miles long and less wide in width, its eastern border is a pristine beach which runs along the Atlantic Ocean. A significant portion of the barrier island is undeveloped, dense with forests and thick undergrowth. Some areas contain former pastures where herds of cattle were raised. A small community of blacks, some of them descendants of slaves, inhabit a small portion of the island's center.

Thomas Spalding, planter, banker, and member of the Georgia House and Senate and of the U.S. Congress during the presidency of Thomas Jefferson, was at one time the owner of Sapelo. Howard Coffin, the developer of Sea Island and much of St. Simons Island, bought the island in 1912 and sold it to R. J. Reynolds in 1933. In 1953 the University of Georgia's Marine Institute was formed with the aid of funding by Reynolds.

Apartments and single-family homes were provided for scientists at the institute and the very large tiled dairy barn was converted to modern laboratories and offices.

The mansion, built originally by Spalding in the Palladian style of architecture by one of my forebears, Reuben King, as contractor, was expanded and refurbished by Coffin and later by Reynolds. The house is one of the most striking buildings in the coastal area. It contains a bowling alley in the basement, an indoor swimming pool, and a large banquet hall. In front a wading pool with Italian statuary reflect the mansion's towering Ionic columns.

It was in the converted barn that members of the Senate Industry and Labor Committee met to be given lectures on marsh ecology and value of the estuary.

This was a red-letter day for me. I'm not sure when or where it happened. It may have been on the boat ride from Sapelo back to the mainland or at the luncheon the members of the Marine Institute had provided. I only know that Senator Holloway confided in me that his father had been an oysterman in the Chesapeake Bay area and that when he (Senator Holloway) was a boy his father's business had been destroyed by pollution of the oyster grounds.

Senator Holloway did not tell me he was in favor of my legislation, but I was sure that his sentiments were in line with my objectives.

The public hearing at Brunswick Junior College was much as I expected. The *Brunswick News* had reported the time and place of the meeting and *Coastal Illustrated* had devoted six columns to Eugenia Price to tell her earlier experience in testifying before the State Institutions and Property Committee in Atlanta and to encourage those in favor of the legislation to attend the interim committee's hearing at the college.

Since a great deal of time has passed since the hearing, I have only two vivid memories of what transpired there. I do know that David Gould, supervisor of the Coastal Fisheries Division of the State Game and Fish Commission, read a letter from George Bagby, the commission's director, encouraging the members of the interim committee to name the commission as protector of the marshlands. Secondly, I recall one of those who testified against the bill. He told the interim committee members that in a chance meeting with me on the grounds of the Glynn County Courthouse, I had informed him that I didn't care if House Bill 212 was constitutional or not, that I planned to ram it through. I could not let that statement go unchallenged and rebutted it in the few remarks I made that day.

According to an account in the *Savannah Morning News* on Sunday, October 19, about 125 people attended the hearing, at which some 25 spoke. I was quoted as saying that I was surprised that no representatives from industry spoke, that some were present but only to take notes. I also told the reporter from the *News* that if the bill were altered to the point where it deluded the people of the state, I would personally ask that it be killed. I predicted that by the end of the 1970 session of the General Assembly a strong marshland protection measure would be passed.

In early November, editorials appeared in several newspapers that for the most part were supportive of marshland protection. On November 2, an article by Jeff Nesmith appeared in the *Atlanta Journal Constitution*'s Sunday edition.

TIDE OF ABUSE HITS CONSERVATION

Reid Harris may not be a member of the Georgia House of Representatives after next year's elections.

From time to time the Brunswick lawyer and State Legislator will usher an old political supporter into his office only to hear the friend inform him that henceforth the support will be elsewhere.

Members of the local Chamber of Commerce are peeved with Harris. So are the labor unions. Brunswick and Glynn County officials have criticized him.

Harris antagonized the local power structure in his home county early this year by introducing a single piece of legislation in the Georgia General Assembly.

The bill with which Harris sent half the important people in his home county running in circles and for which he might sacrifice his seat in the House was designed only to do one thing.

IT WOULD, according to its supporters, have been a timid first step by the State of Georgia to begin providing a measure of protection for the extensive, unspoiled salt marshes of the coast.

It would have declared that marshes are a state resource that must be preserved, a vital link between land and the riches of the ocean, an unmatched recreational treasure.

No other state has estuary resources to match the marshes of Glynn and Chatham and Liberty and McIntosh and Camden, they said.

Harris wanted to create a state board that would have authority to review the plans of big business with regard to the salt marsh, pass on requests from developers to fill large expanses of marsh, pass on requests from developers to build "waterfront" lots, examine the plans of outfits that have expressed interest in stripping away the marsh areas in search of phosphate and other minerals.

THE BOARD, according to big business interests and jealous local politicians, would give the state the power to thwart private enterprise.

At a public hearing on the bill, the first of several, representatives of the county and city governments came to Atlanta and objected loudly that it would usurp their local authority.

The mayor [he should have said the city attorney] of the City of Brunswick and the city manager inserted a telegram into the record, warning, that a requirement in the bill that anyone using marsh area had to first obtain clear title would interfere with the town's urban renewal project.

An urban renewal official later described this contention as "hogwash," because government funds can't be spent on any land until clear titles are established.

Leaders of a local industrial group read a resolution condemning the bill.

Then they hurried home to Brunswick and called a meeting of their people to approve the resolution.

A labor organizer stood in the House Chamber and threatened Harris with political reprisals for introducing a measure that might cost jobs.

If opposition from the home folks wasn't enough to ruin the bill's chances, Harris ran into trouble from his colleagues in the legislature.

Rep. Phillip Candler of Milledgeville, champion of several reform moves in the General Assembly, was chairman of the House committee that reviewed the marshlands bill. He was cool towards it.

Harris' fellow coastal area representatives actively fought the bill and delivered a few hot speeches on the floor of the House when a newspaper story suggested they were being influenced by large paper companies that opposed it.

Rep. Charles Jones of Hinesville was against the bill. He is a lawyer for Interstate Paper Co. according to a national directory of lawyers and their clients.

"We love the marshlands," he said. "I am the fifth link in a chain of title going back to a grant from the King of England. We don't want to decimate the marshlands. We just don't want anyone up there in Atlanta telling us what we can do."

Rep. Dick Scarlett, the second legislator from Brunswick, spoke out against his colleague's proposal. His law firm, according to the directory, represents Brunswick Pulp and Paper Co. and Hercules, Inc.

Rep. Robert Harrison of St. Marys has numerous clients in his law practice, the directory says, clients like St. Marys, Camden County, the City of Folkston and St. Marys Kraft Co., a big pulp interest.

HE WAS AGAINST the bill.

Despite his opposition, Harris managed to get the bill passed, but only after he had taken out the board provisions and simply required that users establish a clear title.

With the Speaker of the House casting the deciding vote, the bill passed and went to the Senate.

It is still in the Senate, where Senator Al Holloway of Brunswick [the author meant Albany] has kept it locked in his Industry and Labor Committee for a year.

Holloway warned that the bill might endanger industrial development in that area.

During the year Holloway and his committee have been waiting, Senator Bob Walling of DeKalb County, a member of Holloway's committee, has introduced Harris' original bill to the Senate.

HE SAYS now that the House-passed bill is so weakened that it is "a deception," something that leaves a comforting impression that it provides protection, but does nothing else.

He will try to get the Senate to pass the original bill.

Meanwhile, worried conversationalists [sic] are wondering if the General Assembly will act in time to provide relief from increasing abuses that "industrial development" in that area is heaping on the salt marshes.

Jeff Nesmith soon followed this article with another *AJC* article headlined "Chatham Phosphate Valued at $16 Billion," which brought about renewed concerns. The article follows:

A phosphate deposit so large that it could propel Georgia into a position as "one of the leading phosphate states in the nation" lies, untouched, beneath salt marshes in Chatham County, state officials said Monday.

The portion of the deposit located beneath a rectangle of saltmarsh and shallow ocean seven miles wide and ten miles long is estimated at a value of $16 billion in a recently-completed Department of Mines study.

The study, which is to be released this week, concludes that about $8 billion worth of the rock is recoverable under present mining and processing techniques.

Much of the land studied is owned by Kerr McGee Corp., the Oklahoma oil company that created a furor in the state a year ago with an application to mine the Georgia coast.

Department of Mines geologist James W. Furlow states that the deposit is "undoubtedly well worth mining" and is probably only the beginning of a belt of rich phosphate extending several miles out to sea.

The 64 square mile area studied in the geological survey by Furlow includes Little Tybee and Cabbage islands, both owned by Kerr McGee; Wilmington Island, under Kerr McGee option, and Wassaw Island, recently purchased by a conservation group and turned over to the U.S. Department of Interior for preservation.

Becomes Thinner

From a series of drillings, Furlow concluded that the phosphate stratum becomes thinner as it reaches inland toward Savannah, deeper as it extends south from Chatham County and closer to the surface to the north toward the South Carolina coast.

Furlow also stated that the rock could be mined without disturbing vital subterranean water resources from which Savannah pumps 65 million gallons of drinking water each day.

And Department of Mines Director Jesse Auvil criticized "preservationists" who have opposed upsetting marshlands in search of minerals or for industrial development.

Sense and Order

"We are trying to get some sense and order and reasonableness started in considerations like this," he said. "Its time the state did things on fact and not fear."

Auvil said that last year's furor over Kerr McGee plans to mine Chatham County marshes was "two and a half days of distorted, bent truths."

"People are just going to have to strike a balance," he said. "Conservation implies using natural resources. It doesn't mean preserving them."

The department of mines is responsible for promoting mining development in Georgia.

Furlow's study states that the Chatham County deposit lies beneath 70 or 80 feet of "overburden" topsoil which usually must be removed along with a thinner strip of clay, to get to the phosphate.

The thickness of the layer of phosphate ore varies from 15 to 60 feet, Furlow said.

Where present consideration is given to mining, in the vicinity of Wilmington, Cabbage and Little Tybee islands, the matrix (strip of ore) is 20 to 30 feet thick, he said.

In addition to the area considered above, evidence indicates that the phosphate matrix may extend continuously out to sea for at least 10 miles and probably beyond the study states.

Thus, the coastal and offshore phosphate resources of Georgia may prove to be extremely valuable as the demand for this mineral commodity increases.

Marine Life Food

Salt marshes, recently found to play an indispensable role in production of food for marine life, can be redeveloped after they are stripped away for mining purposes, Furlow said.

He said "some biologists" believe this can be done in ten years. He said phosphate mining would destroy only about 200 acres of marsh each year.

However, one state conservationist described this idea as "hogwash."

"It takes 10,000 years to build a salt marsh," he said.

When Jeff Nesmith wrote his *AJC* article headlined "Tide of Abuse Hits Conservation," the reporter was not aware that I did not plan to offer for re-election. He probably thought I would be defeated if I offered to run again. My legislative term lasted through 1970. Many people believe that their elected representatives to the General Assembly are paid enough to support their families, make the mortgage payments, and put food on the table and clothes in the closet. This is far from the case. I needed to get back to my law practice to make a living, to be with my wife and our three sons.

My representative district was further from the Capitol than most any other in the state. During my first years of service there was a Pullman car that left Brunswick at ten in the evening and arrived in Atlanta at around seven in the morning. This luxury was soon terminated. Then there were two choices. One could drive at a time when the interstate system was being built, meaning on two lane roads through many small towns. Or one could fly from St. Simons Island to the Atlanta airport on a propeller driven aircraft, later from the Brunswick airport by jet. In six years well over eighty-four trips of three hundred miles each way were made, enough to circle earth at the equator.

During these six years there were times of high drama, but these times were far apart. There were always times of loneliness, of missing my wife and children, particularly at day's end. At times during these summer months, I dreamed of the time when this phase of my life would be over and normalcy would return. No more absences, no more flying through storm-filled clouds, no more winding over narrow country roads. But for now, I knew there was a job to be finished, a job I felt was the most important I had undertaken in life. The marshland legislation must be passed and signed into law.

November passed quickly. The Christmas season ended. The 1970 session of the General Assembly was approaching fast. The deep abyss of coming separation crept into family life. The feeling of separation had not been so intense during the first years of my service, but now it rode through the house like a harbinger of doom.

"Have you really got to go?" one of the boys would ask.

"Yes, I've really got to go."

"How long will you be gone?"

"I'll be gone about a month and a half," I said, trying to reassure.

"That seems like a long, long time to me."

In her encouraging way, the boy's mother would say: "The time will go by quickly." When she was alone with me it was different. "Why do I start missing you before you're gone? This is the worst year of all." She did not cry but tears were brimming.

"Just keep remembering that this will be the last time I have to go."

"I know, I know, I know. But knowing doesn't help me *now*. It doesn't help the big boys *now*. I need your help with the little one *now*. Most of all *I* need you. Sometimes I wish I never heard the word *marshlands*."

"You don't mean that."

"No, I'm sorry. I shouldn't have said that. I'll just be glad when it's all over."

On Monday, January 12, 1970, the legislative session began.

For those unfamiliar with the legislative processes, it may be somewhat difficult to keep up with the convoluted turns the marshlands bill took towards passage. The original bill was passed by committee substitute with the title insurance provisions that I mentioned earlier. In the 1970 session, which met in January, the Senate Industry and Labor Committee determined that the House bill as passed by committee substitute should be changed back to the original version with a few modifications. Therefore the Senate Industry and Labor Committee prepared its own substitute for House Bill 212.

Hearing in the Senate Industry and Labor Committee

Al Holloway, the chairman of the Senate Industry and Labor Committee, had given notice that the committee would have a public hearing on the marshlands bill on February 3, 1970. Again a delegation from Glynn County appeared, led by members of the city of Brunswick's commission, the commissioners of Road and Revenues of Glynn County, and Glynn County's attorney, Edward Liles. Another local attorney, John Gayner, indicated that he was there representing several local industries including Hercules Powder Company, Brunswick Pulp and Paper Company, and Sea Island Company. He said that among his clients were the Brunswick–Glynn County Chamber of Commerce, the Glynn County Committee of 100, the Brunswick Central Labor Council, the Glynn County Real Estate Board, and the Georgia Business and Industry Association. Again, Eugenia Price and Hoyt Brown came to Atlanta from the coast, and Dr. Odum came from Athens to testify. Miss Price dealt once more with the aesthetic values of the marsh landscape, Dr. Odum with its productivity, and Hoyt Brown dealt on the destruction of the marsh that had taken place in the past for the construction of industrial complexes and roads.

John Gayner, a former senator from my home district, stressed the fact that the city of Brunswick is on a peninsula surrounded on three sides by marsh. Under the provisions of the bill, he said, no progress in terms of new industry or development of other kinds could take place in the Brunswick area. (Of course he did not mention the fact that thousands of acres of timberland lay adjacent to the city which could be utilized for growth of many kinds.)

Most of the disgruntled Glynn County delegation left the Senate hearing headed for the Atlanta airport to return to Brunswick. The next day, Wednesday, February 4, 1970, the following article in the *Brunswick News* reported an accident. Some say that the accident was a sign that those opposing the marshlands bill were treading on forbidden ground. Others say that the accident was pure coincidence.

17 RESIDENTS UNHARMED AS PLANE WINDOW BREAKS

Seventeen persons from the Brunswick area, including a delegation of city and county officials, had a narrow escape last night when a window blew out in the pilot's compartment of a Delta airliner.

Traveling at 9000 feet, the plane was brought in for a safe emergency landing at Macon, although the pilot had lost radio contact when equipment was sucked out the broken window. The co-pilot was pulled partially through the window by the sudden, violent vacuum, and suffered a cut arm.

The plane, a Convair 440, took off from Atlanta as Flight 440 at 6:35 p.m., yesterday with 44 passengers, a stewardess, Linda Flannagan, the pilot, Capt. H. F. Kerr, and co-pilot, First Officer J. L. Jones. It was due at its first stop, Savannah at 7:47 p.m. and at St. Simons Island's McKinnon Airport at 8:34 p.m. Another Delta Convair was flown down from Atlanta and took off with the transferred passengers at 10:57 p.m. The Savannah-St. Simons leg was completed at 12:35 a.m.

The delegation, which had gone to Atlanta to attend a hearing of a Senate interim committee on the marshlands protection bill, included Commissioners Hugh D. Leggett, A. L. Wooten and R. L. Holtzendorf, and County Attorney Ed Liles, City Commissioner Thomas E. Morris, and Jim Hall, director of the Chamber of Commerce.

Leggett and Liles were seated in the front left seats, just behind the door leading to a small foyer behind the pilot's cabin.

"It was a close call," Liles said today. "We could feel a sudden vacuum when the window went out. The inner door to the pilot's cabin was pulled off its hinges and flung down inward into the cabin...."

> "The door to the passengers section was pulled inward and Smitty Ledbetter got up and tried to hold the door shut to stabilize the pressure...."
>
> "The pilot did a good job of getting us down without radio to the field in Macon."...
>
> Jones, the co-pilot, was fastened by a seat belt which kept him from being sucked through the window opening. His ear-phones and other movable equipment in the pilot's quarters were reported to have been sucked out of the airplane. Jones was treated for a cut arm in Savannah. No one else suffered injury.

During Gayner's presentation I thought of the fact that he had been dubbed the "constitutional conscience of the Senate" when he served there and that he had been the author of Georgia's Administrative Procedure Law. I recalled that I was the first student that Gayner had met in third grade when his family moved from New Jersey to Brunswick. Much later in high school we had been on the same side in debating whether Macbeth or Lady Macbeth was the greater villain. We worked for hours together for our presentation before the high-school assembly.

Gayner stayed in Atlanta for an extra day speaking with members of the Industry and Labor Committee and other senators and was able to get two paragraphs added to the act's preamble:

> WHEREAS, in the exercise of its police power the State of Georgia recognizes that it is necessary for the economic growth and development of the coastal area that provision be made for the future use of some of the marshlands for industrial and commercial uses; and
>
> WHEREAS, it is the intent of the General Assembly that any use of the marshlands be balanced between the protection of the environment on the one hand and industrial and commercial development of the other.

While John Gayner was buttonholing numerous senators, Drs. Fred Marland and John Hoyt were in conversation with Sen. Al Holloway, having trailed him to his apartment at the Peachtree Towers, explaining and answering questions on one of the state's most valuable resources. They also talked with as many other assemblymen as would listen.

After the Industry and Labor Committee hearing, Eugenia Price returned to St. Simons and wrote the following article for *Coastal Illustrated*. Her description in this essay tracks to a large degree the sentiments she expressed when she testified before the House committee in 1969 and the

Senate committee in 1970. She dealt not with science but with silence, with tranquility, with the aesthetics of the marshscape.

> Just as dawn began to slip over the small marsh behind my house on a chilly January morning of this year, I opened my garage doors to what represented a near miracle to these city deafened ears. I saw nothing but the first sky full of roseate light—a pale blue around the edges as the new wonder broke out of the ocean to the east—and hung above the thick miasma still tucked snugly at the marsh margins, but what I heard I will never forget. I was in a hurry to catch the morning flight to Atlanta, but even if I had missed that plane, I would have stopped long enough to listen to the sudden sounds moving swiftly across the misty marsh that special dawn. All sorts of wrong, city-ignorant thoughts flashed through my mind as I stood there—alligators? Dogs over from Harrington hunting in the marshes? No. The sounds were too loud for dogs and there are no alligators in my marsh anymore—and then I knew what it was I was hearing—soft rapid hoof beats on the black marsh mud, scarcely rustling the dew drenched spartina grasses—thudding, flying away from me, their beat diminishing, fading, but held to my ears by the unbelievable quiet of a marsh at morning—until instant and clear, unmistakably, one, then two, then three deer, invisible in the mist, splashing into the waters of Dunbar Creek and began to swim steadily to the other side.
>
> It had frightened them, opening my old fashioned, quite noisy garage doors so suddenly at a time of day no sound disturbs them from my house on its lonely point of land. And even in their flight they had blessed me. They had given me a set apart moment of the kind of delight man can never give to man. A silence accentuated by their flying hooves and their quick swim across the salt creek—silence I will never forget

⁓

The deer are almost gone from St. Simons Island. My three invisible marsh travelers would not have been there if the marsh had not been there. If it had been filled and built on and dead forever. Man needs silence. It is so rare now to find a part of one minute in which we can still "hear" the silence—even on St. Simons Island. I live in a remote place and in one of those infrequent times when there is not even one plane in the air, not one power saw mutilating a single tree, I stop typing, stop reading, stop whatever I'm doing—to listen. It happens so seldom anymore.

The marshes across which my three dawn visitors fled help hold what silence we still possess in Coastal Georgia.

I was on my way that morning to catch the early plane to fly to Atlanta in order to testify before a Senate Hearing for all of us who still care, in what might be our last chance to save our beloved marshlands—Representative Reid Harris's

Wetlands Protection Act. It seems very strange to most of us that anyone would fight a measure like this, but they are fighting it. Otherwise kind, good people are fighting it.

As I stood listening to the miracle of the deer beside my own little stretch of mist-covered marsh that morning I was glad I had decided to put off my own work and go. Glad and grateful that I could go.

On Friday, February 6, 1970, the Senate passed the marshlands bill by a vote of thirty-nine to zero. The Senate voted for the Senate Industry and Labor Committee's substitute to House Bill 212 without a dissenting vote. I knew that the wires, letters, telephone calls, and buttonholing by the Senators' constituents and many others had convinced them to vote in favor of the bill. After all, Sen. Al Holloway, the committee chairman, had told the press that he had received more correspondence on this one piece of legislation than he had on all other matters before the assembly in his twelve years of service.

To a large degree these letters were from the Garden Clubs of Georgia, Georgia Federation of Women's Clubs, and students and individuals from throughout the state.

As I have said earlier my friend, Jane Yarn, had done an unbelievable job in contacting organizations and individuals asking for support. Her work had paid off. Now the question was, had the House been bombarded with enough facts favorable to the issue to vote in favor of what the Senate had done?

It was fitting that Bob Harrell of the *Atlanta Constitution* wrote of Jane Yarn in the paper's January 8, 1970, edition. Harrell had this to say about a great woman.

MRS. YARN SEES BEAUTY IN THE WORLD

It was a cold rainy night in Thomaston when Mrs. Charles Yarn walked to the front of the room and accepted her award. She was named Conservationist of the Year by the Georgia Sportsmen's Federation and the Sears Roebuck Foundation.

Now you would think a woman would come in a distant second as compared to a man in conservation work. You'd think this because you don't know Mrs. Yarn. When it comes to riding the seas of conservation work, Mrs. Yarn leaves a wide wake. There is no secret to Mrs. Yarn's success in conservation. All it takes is work. All it takes is concern.

Egg Island

Mrs. Yarn first became known to conservationists when she just went out there on the open market and bought Egg Island—with the backing of the Nature Conservancy. This deed put Egg Island beyond the grasp of commercial or industrial developers. Mrs. Yarn found donors who contributed enough to pay back the Nature Conservancy. Egg Island will remain a Georgia coastal gem, a unique and irreplaceable piece of environment.

....

I asked Mrs. Yarn, "What makes you do this sort of thing? There's no pay. Even if there were you'd be underpaid."

In a Shell

Mrs. Yarn considered, then said, "I think all of us are concerned with the world around us, our total environment. But I think some of us might have been submerged too long in our special lives, such as the city, to where a shell or a facade has been formed between us and our total environment. Many of us just can't see our relationship to the world around us."

To talk with Mrs. Yarn is taking a step or two toward establishing contact with the total environment. "Of all the creatures on earth man is the only one who can reason. Man is the only creature who can improve his environment. Who do you think is destroying this environment?"

Mrs. Yarn aims her conservation efforts toward the Georgia coastal areas. Why? "Because this area of Georgia is least tampered with. There is still time to save some of Georgia's coastal islands and the marshes."

How does Mrs. Yarn plan to go about this? She is helping to organize the Planning and Conservation League of Georgia. Mrs. Yarn explained, "I have learned that in all conservation work all roads lead to the gold dome of the Capitol. This is where it happens. The PCL will try to keep up to date on all conservation legislation, dispense it, and lobby for it."

Mrs. Yarn is conservation chairman for the Garden Clubs of Georgia. Politicians might be interested in the fact that these good ladies are 22,000 in number. It is Mrs. Yarn's mission as conservation chairman to interest these women in serving Georgia through conservation. She'll do it, too.

So there you have just one conservationist, Mrs. Charles P. Yarn. It's coincidental, of course, that her husband is a plastic surgeon, a physician dedicated to the rebuilding and making normal the human form. As a conservationist, Mrs. Yarn might be termed a physician of the earth, whose practice is preventive medicine. And although her conservation skills may not be so immediately critical as her husband's, Ms. Yarn's conservation work, as the work of all conservationists, affects our lives—for the better.

After the Senate passed House Bill 212, I returned home for the weekend of February 7 and 8. On reading the local paper, I discovered that on February 4, one day after the Industry and Labor Committee hearing on February 3, the owners of Sea Palms, a resort community on St. Simons Island, within my representative district, appeared before the Brunswick–Glynn County Planning Commission seeking permission to construct a golf course. The area to be utilized was marsh located on the resort's eastern border. A member of the commission, J. D. Compton, a vice-president of the Sea Island Company, suggested that the plans be tabled until such time as final action was taken on House Bill 212. A vote was taken and the plans were tabled.

I also discovered that in late January, a few days before the Senate hearing, the Glynn County Board of Commissioners had approved the filling of twenty-three acres of marsh bordering U.S. Highway 17. At the Senate hearing, testimony in behalf of the commissioners had stressed that the commissioners could handle all matters regarding marshlands, stating that no legislation was needed to protect the marsh. Hoyt Brown of the Marshes of Glynn Society and former member of the zoning board showed in several case studies that local authorities would not protect the marsh.

Before the marshland bill's effective date of July 1, 1970, a condominium was built on a filled portion of the twenty-three-acre marsh tract. Later, a number of units in the complex were severely damaged due to sinking ground. A profusion of lawsuits resulted.

The owners of Sea Palms appealed the ruling of the Planning Commission to the county commission. The commissioners of Roads and Revenues of Glynn County defiantly reversed the Planning Commission's action of putting the golf course plans on the table and granted approval to the filling of the marsh and building the golf course.

The next year, the Sea Palms owners filed an application with the Marshlands Protection Agency. Upon review, the agency denied the application. It is of interest to note that the member of the agency moving that the application be denied was the head of the Georgia Department of Industry and Trade. The nine-hole golf course was later built on high ground, and the marsh that might have been destroyed remains in its pristine state.

The Bill in the House Again

While at home on February 7 and 8, I studied a number of sections of the rules of the House of Representatives.

Back in Atlanta, I sat in the House Chamber early on Monday morning, February 9, 1970, poring over several days of House journals, giving particular attention to matters surrounding the period of unanimous consent.

Finally, I decided to gamble.

When I had finished my research at about 7:30 a.m., I went over to the Senate and paced in front of the door to the office of the secretary of the Senate, waiting for him to arrive.

Within minutes, I spotted him getting off of the elevator.

"Bet I know what you want," he said.

"I want House Bill 212 back over in the House. Do you think you can get it over before the period of unanimous consent has ended?" I asked.

He assured me it would be transmitted about the time the House opened its morning session on Monday, February 9, 1970.

"I hope you get this through the House," he said.

I was surprised at his comment. As a general rule, members of the clerk's staff in the House, the Senate secretary's staff, and members of the Legislative Counsel's Office were hesitant to express themselves on pending legislation.

Back in the House Chamber, I sat in my seat and watched as the Speaker entered, ran down the center aisle and climbed the steps to the rostrum. He motioned to the clerk to activate his microphone.

"Mr. Doorkeeper," he said, "lock the doors and keep them locked."

The doors are locked during the time of morning worship. The Speaker sat down, and a member of the House took his place at the microphone. As I waited anxiously in my place, a representative introduced his pastor in the longest introduction I had heard in all my mornings in the assembly. When the pastor gave his homily, it, too, seemed interminable.

At last the devotional period ended, the Speaker thanked the representative and his guest and instructed the doorkeeper to unlock the doors.

My eyes riveted on the chamber's center doors. No sign of the Senate's messenger. No action could be taken on the bill unless it was physically present in the House.

The chairman of the Committee on Auditing, Enrolling, Engrossing, and Journals reported that the journal of the previous legislative day had been read and found to be correct.

By unanimous consent, the reading of the journal was dispensed with and the journal was confirmed.

The call of the roll was dispensed with.

Still no messenger from the Senate.

In short order, other minor business was transacted.

Unable to sit any longer, I determined to return to the Senate and confer with the secretary again. As I turned into the aisle, a messenger arrived. The business on the floor was now concluded, and the doorkeeper shouted to the Speaker.

"Mr. Speaker, a message from the Senate."

"Let the messenger be received," answered the Speaker.

The messenger walked quickly down the center aisle toward the Speaker's high rostrum, but even as he moved, the membership of the House, not interested in the present proceedings, exploded into motion. Some left their seats, seeking out colleagues. Others went to nearby restrooms and telephones. The hubbub grew, drowning out the voice of the messenger who stood at the small platform in front of the rostrum, speaking from the well.

The messenger turned from the platform, handed the wire carrier by which he had transported the bills to the clerk of the House, bowed lowly to the Speaker, one arm at his waist in front and one arm at the middle of his back, and left the chamber at a trot.

"May I see what the messenger brought?" I asked Glenn Ellard, the clerk of the House.

"Sure, why don't you sit right here and look through the stack," the clerk responded.

I looked at the numbers on the top of several bills. House Bill 212 did not seem to be there. On closer examination, I found what I was looking for buried among the other bills and hurried back to my seat. I stood there near the back of the chamber and waving my arms above my head called out, "Mr. Speaker, Mr. Speaker," my microphone not on.

The Speaker was using his gavel with unusual force, trying to get the membership to settle down.

The Speaker's eyes shifted towards me. He put his gavel down. Several seconds passed and then there was a tapping of the Speaker's pencil point on the high dais, coming loudly through the sound system.

Tap, Tap. Tap.—Tap. Tap. Tap. Tap. Tap. Tap. And finally,

"The gentleman from Glynn is recognized," the Speaker said.

I knew at that instant I could go forward. My study of the house rules and journals had led me to the gamble and the gamble had paid off.

The Speaker had two options when recognizing a House member.

If the Speaker had said, "For what purpose does the gentleman rise?" and I had responded, "For the purpose of having the House agree to the Senate's amendments to House Bill 212," I was certain the Speaker would have responded, "The gentleman is not recognized for that purpose at this time for that motion." If the Speaker had chosen this option, I was dead in the water.

Instead the Speaker chose a second option. He had said, "The gentleman from Glynn is recognized." Under the parliamentary procedures long recognized in the House, I had been granted the floor and could proceed with whatever was on my agenda.

"Mr. Speaker," I called out, "I move that the House agree to the Senate's amendments to House Bill 212."

I heard the Speaker groan. He bowed his head and was quiet for a moment. Then he responded. "Will the gentleman approach the dais?"

I walked down the aisle, climbed the steps, and stood beside the Speaker.

"I had planned to call the supplemental appropriations bill next," he said, "Can't we put this off?"

"Mr. Speaker, you can't possibly know the pressure that I have been under concerning this matter. I simply must get through one way or the other," I said.

"I'll make you a promise," the Speaker said." I'll call your bill right after we finish appropriations."

"I can't wait," I said.

Grasping at straws, knowing that no action could be taken if the bill was not in the chamber, the Speaker turned to the clerk and asked if the bill was in the House.

The clerk replied that it was physically present.

"All right then, here we go," the Speaker said, shaking his head.

As I walked up the long aisle to my desk, I heard the Speaker say, "The motion of the gentleman from Glynn is that the House agree to the Senate amendments to House Bill 212. Is there objection?"

As I passed down the aisle, I noticed the representative from Camden caucusing with some of his allies. They had come over to his desk and pulled out their rule books searching feverishly for some way out. A rule was found that they seemed to like. The gathering dispersed and the Camden representative was standing.

"I object, Mr. Speaker," he said

"The gentleman will state his objection," the Speaker responded.

"The membership of the House has not seen the Senate amendments," my opponent said.

"That's not a proper objection," the Speaker said.

"Well, Mr. Speaker, in light of the fact that the Senate only last Friday passed this bill by committee substitute and in light of the fact that the members of the House have not seen the amended version of the Bill and I am sure would like to know what they are voting on, I move that House Bill 212 as amended by the Senate be printed," the Camden representative said.

"The gentleman has that right," the Speaker's voice boomed out. Then seeing me waving my hand while clutching my microphone the Speaker said, "The gentleman from Glynn is recognized, with the admonition that a motion to print is not debatable."

"Point of inquiry, Mr. Speaker," I said.

"The gentleman will state his inquiry."

"Is it not true, Mr. Speaker, that the Senate merely changed the bill back to the form it had been in when introduced, but changed many features previously objected to by the House?"

"If the gentleman states these matters as facts, the Speaker, having no knowledge of them, has no reason to doubt the truth of your statement. Gentlemen, the appropriations bill is awaiting. Now, the motion is to print House Bill 212 as amended by the Senate. Those in favor vote aye, those opposed will vote no. The clerk will unlock the machine."

The huge voting boards at the sides of the chamber were suddenly alive. Red and green lights appeared. The Speaker waited a few seconds, told the clerk to lock the machine and announced the result.

"On the motion to print the ayes are forty-one and the nays are seventy-eight. The motion is lost." (If the motion to print had passed, the House could not have proceeded at this time with consideration of the marshlands bill. And since the Senate had amended the House bill, it would have been subject to being placed in a conference committee, generally consisting of three members from both houses, and thus further delayed.)

The Speaker moved quickly ahead.

"Now, gentlemen, we are on the motion of the gentleman from Glynn; shall the House agree with the Senate amendments to House Bill 212? The clerk will unlock the machine."

The board lit up again. I stared at the lights trying to determine if 98 were green; the task was impossible. I put my head on my desk and waited for the Speaker's voice.

"Have all the members voted?" the Speaker asked.

"No." A member shouted as he ran in from the back, unlocked his machine, and voted.

"Has everyone voted now?" the Speaker asked again.

There was silence.

"The clerk will lock the machine. On the motion the nays are 21; the ayes are 103. The House has agreed to the Senate amendments to House Bill 212. You can send this to the governor, Mr. Clerk," the Speaker ordered.

I have long cherished the next few moments. A number of the members stood at their seats, turned toward me, some making the victory sign, others applauding and cheering. Wow! What a feeling!!

I moved quickly from the chamber to a bank of telephones and called my wife.

When she answered, I said, "It's over."

"What! What's over?" she asked, puzzled.

"The marshlands bill passed the House. I'll explain it all later," I blurted out.

Almost in tears, she told me how proud she was of me, and I told her how much her help meant to me.

"Let me go. I've got to call Jane Yarn and give her the good news. I love you and we'll celebrate soon," I said.

I called Jane Yarn and told her what happened.

"What? You mean it's really all over?" she said with delight.

"Yes," I said. "This never would have happened without your help."

"I just wish I could have been in the gallery when the last vote was taken," Jane said.

"I wish so too. Let me go now; the supplemental appropriations bill is before the House," I said.

"Give me another minute and explain when the governor will most likely be acting upon the bill," she said.

"As I recall, if the governor calls for a bill while the legislature is still in session, he has five days to act, Sundays excepted, to sign it into law or to use his veto power. If he does nothing with the bill, it becomes law in the five-day period. If the bill is sent to the governor after the end of the session, the period is thirty days, Sundays excepted. If he does nothing within that time frame it becomes law. He must exercise his veto power in the prescribed period," I recited.

"What's your best guess as to when the bill will be on the governor's desk?" Jane asked.

"I have no way of knowing. I think a campaign to convince the governor that the bill should become law should begin immediately," I urged.

"Your wish is my command," Jane said. "I'll go to work on it right away."

The Bill in the Governor's Office

Now the time for waiting had arrived. The session of the assembly concluded sine die (final adjournment) on February 21, 1970. The bill had not been sent to the governor. Twelve days had passed and the bill still reposed in the Clerk's Office.

Sometime in late February or early March, I received in the mail a composite status report marked "Final Three." The report from the clerk of the House contained a record of bills and resolutions introduced in the House and Senate in 1969 and 1970 showing the fate of each and the dates on which those bills and resolutions that had passed were sent to the governor's office. "Final Three" showed that House Bill 212 had gone to the governor on February 25, sixteen days after passage.

I studied the calendar and determined that Governor Maddox had until March 27 to veto or to sign the bill into law. If he did nothing by midnight on March 27, the bill would become law by terms of the Georgia Constitution.

Time crept by. There was no word on what the governor planned. Then, toward the last of March, a newspaper article reported that the governor indicated that he was thinking of a veto. I called the reporter, who confirmed what he had written.

"The governor thinks the bill is much too controversial," he said.

I immediately called George Bagby at Game and Fish and told him about the article and my conversation with the reporter.

"I heard about that and planned to call you," Bagby said.

"Listen," I said. "You are without a doubt closer to the governor than anyone else favoring this legislation. If you can meet with him, tell him that the bill is one that the Senate's Industry and Labor helped write. Tell him that the composition of the Marshland Agency was made in the main by Sen. Al Holloway. Move the bill away from me and to the Senate."

"You think the governor has some crow to pick with you?" Bagby asked.

"Could be," I speculated. "After all, when the governor's election was left up to the House and Senate, I voted the way a majority of my constituents voted and that was not for Maddox. This I know though, Maddox signed various other bills I authored including the Georgia Surface Mining Land Use Act. Maybe I'm being paranoid. I'm sure that there have been many letters from coastal interests, industry, land developers, and others telling him that the bill will stifle growth in the coastal region, but there is favorable correspondence too."

"I'll see what I can do," Bagby said.

"And will you try to find out when the governor will act on this?" I added.

"I'll call if and when I find out," Bagby answered.

I made several other calls, one to Jane Yarn to make sure she knew what the governor had said, one to Rock Howard at Water Quality, and one to Ben Fortson, the secretary of state. Fortson said the governor put little stock in anything he could say, but assured me that George Bagby had a great influence on Lester Maddox. Jane Yarn promised to get in touch with the Institute of Ecology in Athens and others asking that the governor be contacted as soon as possible. Howard said he would do what he could. I then contacted Dr. Frederick C. Marland at the UGA Marine Institute on Sapelo Island and was told that he and a number of his associates would be writing and wiring the governor.

Now the wait seemed interminable. Days passed with no word from Game and Fish.

Then, on March 26, as pessimism was crowding my thoughts, I got a call at home from Zell Miller, the governor's executive secretary.

"The governor asked me to call you and let you know he will be taking action on your House Bill 212 tomorrow. He would like for you to be present in his offices at the Capitol. Think you can make it?" Miller asked.

"I can be there," I answered. "Does this mean he is going to approve the bill?"

"I really can't say. He's still undecided. I'm going to call Mrs. Yarn, Dr. Odum, and Sen. Holloway. Anyone else you want there?"

"George Bagby, by all means," I said.

"I've already notified George Bagby," Miller told me.

"See you tomorrow then," I said.

The next morning I flew to Atlanta on an early flight and, reaching the Capitol some time before 9:00, went directly to the governor's reception room.

Soon, Zell Miller joined me, and then one by one, Jane Yarn, Dr. Eugene Odum, and Rock Howard arrived. Al Holloway's secretary came in and announced that he had a death in the family and would not be able to come. An alarm bell sounded in my head. Sen. Holloway could have had a great influence on the governor if he had decided on a veto.

"Where is George Bagby?" I asked Miller.

"He's already in there with the governor," he responded.

Time moved slowly as we waited. Finally the governor's receptionist opened her door, and we were told that the governor was ready to see us. We filed through the receptionist's outer office and into the governor's office.

"Come in, come in," Governor Maddox called.

We visitors stood before the governor's desk.

"Nice to see you, little lady," Maddox said to Jane Yarn, and then to me, "I see you made it, Chief." He nodded to Howard.

George Bagby, standing at one end of the governor's desk, looked at me and shrugged as though he had no clue as to what the governor's decision would be.

"Someone seems to be missing," the governor said.

Al Holloway's secretary, Farris Freeman, spoke up.

"Perhaps you mean Sen. Holloway. He had a death in the family that prevented him from coming. He really wanted to be here," Freeman said.

"Too bad, too bad. Well, yes-sir-ree-bob. We've got us a controversial bill here. You know it's controversial, don't you, Chief?" the governor said, looking at me.

"I know, Governor. Nobody in the state knows better than I do."

"Tell you the truth, Chief, I've thought long and hard about a veto. Bagby here tells me I should sign it into law. You ought to see the letters and wires I've gotten on this. There must be three barrels full back there," the governor said turning to his secretary.

"How many pens you got for me?" he asked.

"Here you are, Governor," she said, handing him a handful of felt-tipped pens.

"Here we go then," Maddox said. The governor sat at his desk and picked up one of the pens.

"Point me to the right line to make this a law," he said to George Bagby. "Don't want to sign the wrong one and make it a veto."

Bagby pointed to the proper place and the governor commenced to sign.

"Les," he wrote and motioned for me to take the pen. Then he wrote "ter." "Who wants this one?" Finally, using several other pens, he finished writing "Maddox."

"Well, it's done," Maddox said. "Hope I've done the right thing."

"I'm sure you did, Governor," was all that I could say.

Epilogue

After the passage of the Coastal Marshlands Protection Act of 1970, Dr. Fred Marland and his colleagues, Drs. Tom Linton, Jim Henry, and John Hoyt continued their scientific studies at the University of Georgia Marine Institute on Sapelo Island. A year later Marland was selected among several candidates to the staff of the Marsh Agency. He served for twenty-two years, reviewing applications, interfacing with applicants, monitoring the coast, and making recommendations to the Coastal Marshlands Protection Committee. Marland was joined in these efforts in 1975 by his colleague Dr. John R. Bozeman, who also became the principal author of the Shore Protection Act of 1979. That act was sponsored by Rep. Dean Auten of Brunswick.

Jane Yarn was named Atlanta's Woman of the Year. She served on Governor Carter's Environmental Council and later as one of the members of the Council for Environmental Quality, serving as a consultant to President Carter at the White House. A fifteen-thousand-square-foot Interpretive Center was named in her honor at Tallulah Falls, Georgia.

Reid Harris was given the Legislative Conservation Award by the Georgia Sportsmen's Federation and the National Wildlife Federation. Governor Jimmy Carter appointed him to the Governor's Goals for Georgia advisory group and later as the chairman of the Georgia Environmental Council. In 1998, Governor Zell Miller presented Harris with a resolution commendatory of the marshlands act and noted that the law had withstood the passage of time.

Eugenia Price continued to write. She finished her St. Simons Trilogy with *Lighthouse* in 1972, her Florida Trilogy in 1980, her Savannah Quartet in 1989, and her Georgia Trilogy in 1995. Beside this output, she authored more than eighteen other books before her death in 1996.

Dr. Eugene Odum was elected to the National Academy of Science and was named an honorary member of the British Ecological Society. With his brother, Howard W. Odum, he received the Institut de la vie prize. The $150,000 Tyler Ecology Award was presented to Dr. Odum by President Jimmy Carter in the White House. Dr. Odum died in 2002.

In June 1971, the United States Corps of Engineers commenced filling the marsh adjacent to the St. Simons Island causeway. As chairman of the Governor's Environmental Council, Harris contacted Governor Jimmy Carter as soon as he saw the dredging equipment at work. Arrangements were made for a meeting with the governor at the Governor's Mansion. After that meeting Governor Carter issued the following press release:

> Late Saturday evening I received word that dredging operations had begun on a twenty-four-hour-a-day basis in Terry Creek which is immediately adjacent to the Torras causeway between Brunswick and St. Simons Island. Spoil from this dredging operation was being dumped on marshland in ever closer proximity to this causeway.
>
> The most direct result of the continuance of this operation would have been the destruction of 150 acres of scenic marshland immediately adjacent to this causeway. In addition, runoff from the spoilage, which contains a high level of the chemical toxaphene, would have seriously threatened the waters of the marsh.
>
> After consultation with environmental experts both in and out of the state government, I called Col. Strohecker of the U.S. Corps of Engineers in Savannah—at 2:30 Sunday morning. I requested that the dredging operations be halted, which Col. Strohecker agreed to do. Operations ceased at 6:00 a.m. Sunday.
>
> Sunday evening, I met with representatives of the Corps of Engineers, the State Game and Fish Commission, the Attorney General's office, and interested parties from Brunswick and Glynn County.
>
> I pointed out that under our interpretation of the Federal Fish and Wildlife Coordination Act notification and consultation with the State Game and Fish Commission was required. This should have been done some 14 months prior when the initial decisions in the project were made. This was not done. I also pointed out that the Attorney General had ruled that our marshes are state property. Based on these facts, I requested that dredging operations in Terry Creek be suspended until a thorough study could be made to determine the most advantageous method for disposing of the spoilage. The Corps of Engi-

neers readily agreed to this request and stated that no additional dredging in Terry Creek would take place for at least a year. The Corps of Engineers also agreed to allow a representative of the State Game and Fish Department to review plans for dredging of the Inter-coastal Waterway north of Brunswick, with special attention to be given to the proposed sites for dumping of spoil from this operation.

I would like to point out that the U.S. Corps of Engineers has been completely cooperative throughout. I appreciate their goodwill and think they deserve favorable recognition for it.

This is a very complex question with implications that extend far beyond Terry Creek or the Torras causeway. As Governor I am responsible for industrial development as well as conservation.

I believe that it is possible to provide for the preservation of our natural resources and at the same time continue to attract new industry and prosperity to Georgia. However, too often in the past decisions have been made and actions taken on a very narrow and short-sighted economic basis. Little consideration has been given to the recreational value of the area affected or the long-term environmental impact of the action taken. The case of Terry Creek is just one example.

Already the sport and food fishing industry along the Georgia Coast has been seriously affected [by] the cumulative impact of a series of decisions—all of which could be justified in narrow, short-term economic terms.

Even though the area of marshland directly involved in filling, dredging and the dumping of spoilage has been small when compared to the one-half million acres in the marsh, the consequences of these actions have extended to almost every sector of the marsh.

Acre for acre, the Georgia sea marshes are some of the most productive real estate on the face of the earth.

The Marshes of Glynn belong to the four and one half million people of Georgia. If we allow them to be destroyed, all the technology and laws on earth will not be able to replace them.

As elected representative of the four and one half million owners of the marshes, I intend to do what I can to see that our children will not have to face that day.

Thirty-seven years have passed since the passage of the Coastal Marshlands Protection Act. A number of states have used portions of the act in the passage of marsh protection legislation. Appellate courts in Georgia have not found the act to be unconstitutional.

The Supreme Court of Georgia, in an opinion by Justice George Carley, decided unanimously on May 17, 2006, *Black v. Floyd*, 280 Ga. 525, held that the state of Georgia continues to hold title to the beds of all tidewaters within the state, except where title in a private party can be traced to a valid Crown grant or state grant which explicitly conveyed the beds of such tidewaters.

Earlier in 1976, the Georgia Supreme Court ruled in the case of *The State of Georgia v. Ashmore*, 236 Ga. 401, the state has fee simple title to the foreshore in all navigable tidewaters.

Some courts have enforced provisions of the law to the extent that violations have been found and remedies invoked. Some who have filled the marsh have been required by the courts to remove the fill.

At the present time, debate is taking place regarding the scope of the law. Some favor extending the reach of the law to such matters as runoff occasioned by upland development. Other experts say that the operation of the law should be confined to activities taking place within the marsh environment. They argue that matters occurring outside the marsh environment—for example, ditching of the upland resulting in runoff—should be controlled by other laws such as the Coastal Zone Management Act.

Several individuals have claimed authorship of the marshlands act. Only four people were responsible for drafting and rewriting it. Those four are Dr. Fred Marland, Virlyn Slayton of the Legislative Counsel's Office, Sen. Al Holloway, and the author.

A number of people have asked the author when he first became interested in the marsh. He has answered by relating the following: The temperature stood in the high nineties in the small coastal town of Brunswick, Georgia. It was mid-July and there were no clouds in the sky. The tide was running out of the creeks when the 1935 black Ford backed out from the porte cochere. An eight-year-old boy sat on the left front fender. His fourteen-year-old brother stood on the right running board. When the car stopped to move north on Carpenter Street, the fourteen-year-old moved to the front right fender and waved to his aunt, the driver, to move on. The car moved northward to Albermarle Street and then turned east.

It passed Wolfe, Albany, Amherst, and Cochran Streets and came to the town's easternmost drive, the Boulevard. The aunt drove south a quarter of a mile and parked on the roadway's left margin. The aunt jumped quickly from the car and the older boy moved from the fender, leaving the younger boy behind. The twosome ran along some wooden planks some two hundred feet to a small dock at the creek's edge and dove into the water.

Running after his brother and aunt, the younger boy saw the swimmers as they approached a curve in the creek and disappeared from view.

The boy jumped in, trying to follow. He sank and fought his way back up. A second time he went down, coughing and sputtering. He tried to move back to the dock, but the tide was too strong. As he went down the third time he felt a strong hand pushing him back up, toward the marsh. Up the mud bank he went, panting for breath and shaking with fear. He grasped at the blades of marsh and pulled himself to safety.

An unknown helper with the aid of the marsh had saved his life. Thirty years later he was fighting to save the marsh.

Afterword

In the 1960s, Reid W. Harris commuted from his St. Simons Island home to his law office in Brunswick, Georgia. His route across a four-mile elevated causeway afforded him a view as far as the eye could see of pristine marshlands. During that same period, I made a nearly identical trip regularly in a school bus, and I make that commute today. The marsh's beauty strikes me every time I cross that causeway. But the security and well-being of the marsh was not always a sure thing, as you have read in this book. Approaching Brunswick on the causeway, there are still clear remnants of the imminent threats to our estuaries during that period. I remember as a student being struck by garish changes to the landscape: marsh being filled using dredge spoils for new residential development extending thousands of feet from the mainland.

Thankfully, we had Mr. Harris to advocate for the rather simple and powerful idea that the expansive Georgia marshes are a treasure shared by all who visit or live on our coast. Today we appreciate how important it was for Mr. Harris to shepherd the Georgia Coastal Marshlands Protection Act (GCMPA) through the Georgia Legislature. The act has been proven legally and scientifically sound and provides groundbreaking environmental protections for this vital habitat. Additionally, the political and environmental communities that worked with Mr. Harris have proven just as resilient as his work on the marsh protection act. Today's current coastal conservation efforts are extensions of work begun during Harris's time in office.

Fortunately for Georgia, Reid Harris was the right person at the right place at the right time. He decided to take deliberate actions to protect vital environmental resources. The Georgia political system—the right place—was in a state of transition, which allowed the legislature to pass effective regulations governing the protection and use of Georgia's coastal ecosystem. At the same time, Harris—the right person—spurred his colleagues to overcome rising industrial and real estate development pressures. All of this occurred in the late 1960s and early 1970s—the right time—a moment of great social, racial, and electoral upheaval, which also saw the nascent environmental movement form around key issues. The political and social context that allowed passage of the GCMPA included the narrow election of the often controversial Georgia governor Lester Maddox. This upheaval and transition gave rise to what has been fifty years of relatively constant political and regulatory enforcement of marsh protection measures.

By the time Reid Harris published the first edition of this book in 2008, he had received general recognition for his work's significance. Georgia's indispensable marsh estuary system had been conserved intact and functional, which cannot be said for other areas of the East Coast that have suffered more extensive degradation through filling, channelization, and other impacts. GCMPA was an early entrant into the small suite of coastal protections that form the basis for the regulatory framework that protects this most biodiverse part of Georgia.

While seemingly the perfect candidate for this effort, Reid Harris did not enter his political career intending to lead this battle. Once elected, Mr. Harris developed a sophisticated understanding of the developing science of ecology to augment his legal discernment on state ownership establishing a basis for regulation in the water bottoms and marshes.

Reid Harris activated an interesting network of support for the GCMPA legislation. He chose his allies carefully, tapping into the organizing power of groups not associated with the traditional and predominantly male power brokers of the day. Throughout the legislative process, he worked with both state and local organizations typically associated with women, such as the League of Women Voters and various garden clubs, to leverage influence at key moments during the GCMPA's journey to passage. Many of the women involved in these groups went on to lead the fledging environmental movement in the state.

Two other figures were key in backing Mr. Harris's successful campaign. Jane Hurt Yarn and Fred Marland prepared him well from the political and scientific standpoint. In 1969 Ms. Yarn was an emerging leader in advocating for coastal conservation issues. Meanwhile the insights from developing science, which had started in the 1950s at the University of Georgia Marine

Institute on Sapelo Island, helped create a compelling need for protecting the marshes and estuaries. The body of work provided by the scientists mentioned in this book provided a sound footing for the extensive research and regulatory framework that continues to this day on the Georgia coast.

The Georgia Coastal Marshlands Protection Act has endured from a legal standpoint, standing up to the varied challenges by providing key protections that have remained in place for fifty years. Much of the credit for the continuity for regulatory consistency rests with what has become known as the Coastal Resources Division of the Department of Natural Resources. Though Mr. Harris had less to do with the implementation and enforcement of his legislation, its staying power provides insight into his intelligence, values, and reputation among his peers.

In an interview near the time of this book's first publication, Reid Harris stated that it took forty years for him to receive honors and appreciation for the passage of the GCMPA. In that time frame, it has become a given that protection of the marshes is held as a priority in local, state, and federal arenas. For the last few years, it has been a challenge to buy or borrow this book, as it has been out of print. Its continued local popularity in coastal Georgia is a testament to both the recognition of Mr. Harris's work and the complexity of the legislation. When I began working with the Georgia Conservancy in 2015, I was told (and took to heart) that *And the Coastland Waits* was required reading for those working in coastal conservation.

Since 1970, the benefits, function, and value of the marsh ecosystem have become broadly accepted and the legal precedent that supports the GCMPA has evolved to govern additional environmental legal questions, including ownership of marshlands, construction of docks, and feasible development on hammock islands. This is largely why Georgia has an intact marsh ecosystem with one of the most functional estuaries in the eastern United States. As the value of adjacent high ground rises, the most significant challenges our marshes face will not be legal challenges but will remain political.

Mr. Harris was a brilliant attorney who brought together legal creativity and political skill to produce policy that will serve coastal conservation well for the next fifty years and beyond. As we pass the fiftieth anniversary of marsh protection, we should consider what threats we will face before the one-hundred-year mark. Sea level rise, in particular, is a harrowing threat, which was not widely understood a half-century ago and is only now entering day-to-day discussion.

In Glynn County, Reid Harris's coastal home county, and elsewhere along the coast, there is now more momentum behind understanding the impacts of sea level rise and its related drivers of climate change. Fifty years from now, the current predictions of rising tides will have come to pass in some measure. The marsh will adapt and change as it has for millennia. Humans will adapt or retreat, but to date there has been little public discourse on the subject. This is the next great challenge, no less significant than what Mr. Harris faced in the 1970s.

I agree with something I heard Reid Harris say in an interview many years ago and I paraphrase a bit: "The marshes are a glorious thing that God has granted to all of us." Mr. Harris's legacy is that we must remain ever vigilant, for there are still vulnerabilities facing this wondrous resource.

Charles H. McMillan III
Coastal Director, Georgia Conservancy
St. Simons Island, Georgia

Appendix A

The Attorney General's Opinion

I was not aware at the time that Governor Maddox signed the marshlands bill on March 27, 1970, that an opinion regarding state ownership of the coastal marshes had been issued by the attorney general's office approximately ten days earlier. It was sent to Col. John Egbert of Savannah District Corps of Engineers. A copy of the letter with the opinion attached was sent to Governor Maddox. Had this opinion been issued a little more than a year earlier, my struggle in getting the marshland bill passed would have been much less difficult. I do not know if Governor Maddox read the opinion or if it influenced his decision to sign the bill into law. I do know this opinion, which appears below, is a document that should be studied by anyone interested in Georgia's claim to ownership of the coastal marshlands.

Position Paper Relating to the Georgia Coastal Marshlands: Written by Courtney W. Stanton and Approved and Issued on March 17, 1970, by Attorney General Arthur K. Bolton.

There have been a number of recent proposals made by firms and individuals relative to the development of coastal marshlands. Since most of such proposals have been made upon the assumption of private ownership of such marshes, the Attorney General thinks it proper to assert the position of the State in this matter. In brief, it is that the marshlands of Georgia are not susceptible to private exploitation or conservation without regard to the common-law trust purposes to which these lands have long been dedicated.

Marshlands, like all lands within the British Empire, were originally the exclusive property of the Crown. The similarity between marshlands and other realty ends here, however, for while land above high water mark was customarily granted to private owners, the sea, its arms, and lands subject to be covered by tides were of such great importance to the general public that it was considered that the Crown held them in common-law trust for the benefit of all subjects. See *Shively v. Bowlby*, 152 U.S. 1 (1894); *Concord Mfg. Co. v. Robertson*, 25A. 718, 720–727 (N.H. 1890). Under the English common-law, the landward boundary of the trust property is prima facie the high tide line.

After the American Revolution, the original States succeeded to Crown properties and also to the Crown's status as trustee, a status recognized by the Supreme Court of the United States. *Martin v. Waddell*, 41 U.S. 367(1842). Such succession and status are also recognized by the Georgia courts. *Young v. Harrison*, 6 Ga. 130 (1849); *Johnson v. State*, 114 Ga. 790 (1902).

The question of navigability of waters does not address itself to the marshlands problem. By earlier English doctrine the matters of the flow and navigability were tightly interwoven, but this was because of a geographical coincidence: all navigable waters in England are subject to the tides, a situation which does not obtain in North America. The American Courts have recognized that tide flow and navigability are not to be confused. *The Genessee Chief*, 53 U.S. 443 (1851).

Ga. Code Ann 85-1304, which provides that "The rights of the owner of lands adjacent to navigable streams extend to the low-water mark in the bed of the stream" does not affect marshlands. Such Code Section, being in derogation of the common law, must be strictly construed, and a marshland is in no sense a "Stream."

Ga. Laws 1902, p. 108 (Ga. Code Ann. 85-1307 through 85-1309) attempted to extend the boundaries of owners abutting tidal waters to the low water mark. Such act, however, was clearly unconstitutional as violative of Georgia Constitution 1877, Art. VII, Par. 1 (Ga. Code, 1933, 2-6401) forbidding legislative grants. For same provision see Ga. Constitution 1945, Art. VII Sec. XVI, Par. 1 (Ga. Code 1933 2-5402 Par. I). The Constitution of 1945, Art. 1, Sec. VI, Par. I (Ga. Code Ann. 2-601) purported to "confirm" the Act of 1902, but such confirmation is of little or no effect.

In the first place, a statute void at its inception does not become valid because of subsequent constitutional action. In the second place, the 1945 constitutional provision does not, per se, contain any language constituting a grant of property, and even if such provision be construed as a grant to the General Assembly of power to enact legislation similar to that of 1902, such power has not been exercised. Even if susceptible to construction as an original grant of title, the constitutional provision was ineffective in that the language failed to express the donative, public intention with sufficient clarity to alert the electors that they are being called upon to vote a grant of public rights to private persons. See *Goolsby v Stephens*, 155 Ga. 529, 540 (1923).

Even if Ga. Laws 1902, p. 108, be taken at face value, it is doubtful that the grant therein contained to the "beds of all tidewaters," to "land adjacent to navigable tidewaters" are sufficiently definite to cover in whole or in part lands which are defined by the U.S. Naval Oceanographic Office Navigation Directory (2nd ed., 1969) as "flatland periodically flooded by salt water." The same publication defines "tidewater" as water affected by tides or sometimes that part of it which covers the tideland.

In the unlikely event that one should establish a title to marshland, such person could not use the property in such a way as to impede the public right of enjoyment thereof unless the grant to the marshland expresses a full relinquishment of all public rights.

Appendix B

House Bill 212

INDUSTRY AND LABOR COMMITTEE SUBSTITUTE TO H.B. 212 FINAL VERSION AS PASSED BY THE SENATE ON FEBRUARY 6, 1970 AND BY THE HOUSE ON FEBRUARY 9, 1970. SIGNED INTO LAW MARCH 27, 1970

A BILL
TO BE ENTITLED

An Act to create the "Coastal Marshlands Protection Agency"; to provide a short title; to define certain terms; to provide for the membership of the Agency; to provide for the election of a Chairman; to provide for the appointment of an executive Secretary; and of representatives and agents by designated members; to provide for payment of expenses; to provide for the powers and duties of the Agency; to provide for applications to alter marshlands; to provide for the procedure for filing applications; to provide for gathering of information by Agency members; to provide for issuance of permits; to provide for conditional permits; to provide for denial of permits under certain conditions; to provide for appeals; to provide for policing; to provide for injunctions; to provide for posting permits; to provide for the transfer of permits; to provide penalties for violations; to provide for exceptions; to provide for emergency powers; to provide for severability; to repeal conflicting laws; and for other purposes.

WHEREAS, scientific research has established that the estuarine area of Georgia is the habitat of many species of marine life and wildlife, and without the food supplied by the marshlands, such marine life and wildlife cannot survive; and

WHEREAS, intensive marine research has revealed that the estuarine marshlands of coastal Georgia are among the richest providers of nutrients to the world, and

WHEREAS, the marshlands of Georgia provide a great buffer against flooding and erosion, and help control and disseminate many pollutants, and

WHEREAS, the estuarine areas and coastal marshlands provide a unique form of outdoor recreation for the people of our State, and

WHEREAS, it is in the public interest that the State of Georgia regulate the use of the coastal marshlands by the exercise of its police power in order to protect the welfare, health, and safety of the citizens of this State; and

WHEREAS, in the exercise of this police power the State of Georgia recognizes that it is necessary for the economic growth and development of the coastal area that the provisions be made for the future use of some of the marshlands for industrial and commercial purposes; and

WHEREAS, it is the intent of the General Assembly that any use of the marshlands be balanced between the protection of the environment on one hand and industrial and commercial development on the other.

NOW THEREFORE, BE IT ENACTED BY THE GENERAL ASSEMBLY OF GEORGIA

Section 1. Short Title—This Act shall be known and may be cited as the "Coastal Marshlands Protection Act of 1970"

Section 2. Definitions—unless clearly indicated otherwise by context, the following terms, when used in this Act shall have the meanings respectively ascribed to them in this section:

> (a) "Coastal Marshlands" hereinafter refers to as 'marshlands' means any marshland or salt marsh in the State of Georgia, within the estuarine area of the State, whether or not the tide waters reach the littoral areas through the natural or artificial water courses. Marshlands shall include those areas upon which grow one, but not necessarily all of the following: salt marsh grass (*Spartina alterniflora*), black grass (*Juncus gerardi*), high tide bush (*Iva frutescens* var. *oraria*). The occurrence and extent of salt

marsh peat at the undisturbed surface shall be deemed to be conclusive evidence of the extent of a salt marsh or part thereof.

(b) "Estuarine area"—means all tidally-influenced waters, marshes and marshlands lying within a tide-elevation range from five and six tenths (5,6) feet above mean tide level and below.

(c.) "Person"—means any individual, partnership, corporation, municipal corporation, county, association, public or private authority, and shall include the state of Georgia, its political subdivisions, and all its departments, boards, bureaus, commissions of other agencies, unless specifically exempted by the provisions of this act.

(d) "Applicant"—means any person who files an application under the provisions of this act.

(e) "Political subdivision" means the governing authority of a county or a municipality in which the marshlands to be affected or any part thereof are located.

(f) "Agency"—means the Coastal Marshlands Protection Agency

Section 3. Creation of the Coastal Marshlands Protection Agency.—

(a) There is hereby created, an autonomous division of the State Game and Fish Commission, the Coastal Marshlands Protection Agency which shall administer the provisions of this Act.

(b) The Agency shall be composed of seven (7) members as follows:

1. Director of the State Game and Fish Commission
2. The Executive Director of the Ocean Science Center of the Atlantic
3. The Executive Secretary of the Water Quality Control Board
4. The Director of the Coastal Area Planning and Development Commission
5. The Executive Director of the Georgia Ports Authority
6. A Director of the Georgia Natural Areas Council to be elected by a majority of the directors thereof.
7. The Attorney General or their appointed representatives. In the event one of the members of the Agency designated herein appoints a representative, such representative shall be an employee of

the same State agency or department as the official making the appointment.

(c.) A majority of the members of the Agency shall elect a chairman from among the members who shall serve for a period of four years from the date of his election and until his successor is elected.

(d) The members of the Agency shall receive no compensation for their services, but shall be entitled to receive actual expenses incurred in the performance of their duties from the agency or department with which employed.

Section 4. Powers and Duties of the Agency—

(a) The Agency shall have the following powers and duties:

(1) To promulgate such rules and regulation as may be necessary to effectuate the provisions of this Act; provided, however, that such rules and regulation shall be of no force and effect unless two public hearings be held after notice thereof has been published in the legal organ in the counties of Camden, Glynn, McIntosh, Liberty, Chatham and Bryan once a week for two consecutive weeks immediately prior to such hearing.

(2) To administer and enforce the provisions of this Act and all rules, regulations, and orders promulgated there under.

(3) To examine and pass upon applications to alter marshlands.

(4) To revoke permits of applicants who fail to carry out their proposals.

(5) To accept monies that are available from government units and private organizations.

(6) To institute and prosecute all such court actions as may be necessary to obtain the enforcement of any order issued by the Agency in carrying out the provisions of the act.

(7) To exercise all incidental powers necessary to carry out the purposes of this Act.

(b) The above and foregoing powers may, except for the rule making power, be exercised and duties performed by the Agency through such duly authorized agents and employees as it deems necessary and proper including an executive secretary.

Section 5. Applications, procedure.—

(a) No person shall remove, fill, dredge or drain or otherwise alter any marshlands in this State within the estuarine area thereof without first obtaining a permit from the Coastal Marshlands Protection Agency.

(b) Each application for such permit shall be filed with the State Game and Fish Commission and shall include:

(1) Name and address of the applicant.

(2) A plan or drawing showing the applicant's proposal and the manner or method by which such proposal shall be accomplished.

(3) A plat of the area in which the proposed work will take place.

(4) A copy of the deed or other instrument under which the applicant claims title to the property, or if the applicant is not the owner, then a copy of the deed or other instrument under which the owner claims title together with the written permission from the owner to carry out the project on his land. In lieu of a deed or other instrument referred to in this paragraph 3 the Agency may accept some other reasonable evidence of ownership of the property in question or other lawful authority to make use of the property.

(5) A list of all adjoining landowners together with such owners' addresses. If ownership of adjoining landowners cannot be determined or if addresses cannot be ascertained, the applicant shall file in lieu thereof a sworn affidavit that a diligent search has been made but that the applicant was not able to ascertain the owners or addresses as the case may be of adjoining landowners.

(6) A certification from the local governing authority(s) of the political subdivision(s) in which the property is located stating that the applicant's proposal is not violative of any zoning laws, ordinance or other local restriction which may be applicable thereto. If in the judgment of the Agency a zoning permit is not needed prior to considering the application, it may waive this requirement and issue a conditional permit based upon the condition that the applicant acquire and forward a permit from the local political subdivision prior to the commencement of work. No work shall commence until this requirement is fulfilled.

(7) A certified check or money order in the amount of $25.00 for each acre of land or portion thereof to be affected payable to the Coastal Marshlands Protection Agency to defray administrative costs. No applicant shall be required to pay in excess of $500.00 for any one proposal regardless of the amount of acres to be affected.

(c) A copy of each application for a permit shall be delivered to each member of the Agency within seven days from receipt thereof.

(d) The Director of the State Game and Fish Commission within thirty days of receipt of an application shall notify in writing all adjourning land owners of the application and shall indicate the use the applicant proposes to make of the property. Should the applicant indicate that any adjoining landowner is unknown or that the address of such landowner is unknown, then the member of the Agency to which the application for permit is filed shall cause a notice of the proposed activity and a brief description of the land to be affected to be published in a legal organ of the county or counties in which said land lies within thirty days of receipt of the application.

Should the property to be affected by applicant be bordered on any side or on more than one side by other property of applicant, applicant shall supply the names and addresses of the nearest landowners other than the applicant and bordering on applicant's land or a sworn statement of diligent search as provided above in this Act. The landowner so named shall be notified either directly or by advertisement as provided above in this Section. Any member may also make inquiry to adjoining landowners to ascertain whether or not there is objection to issuance of a permit.

(e) In passing upon the application for permit, the Agency shall consider the public interest which, for the purposes of this Act, shall be deemed to be the following considerations:

(1) Whether or not any unreasonably harmful obstruction to or alteration of the natural flow of navigable water within such area will arise as a result of the proposal.

(2) Whether or not any unreasonably harmful or increased erosion, shoaling of channels, or stagnant areas of water will be created to such extent as to be contrary to the public interest.

(3) Whether or not the granting of a permit and the completion of the applicant's proposal will unreasonably interfere with the conservation of fish, shrimp, crabs, and clams or any marine life or wildlife or other natural resources, including but not limited to, water and oxygen supply to such an extent to be contrary to the public interest.

(f) If the Agency finds that the application is not contrary to the public interest as heretofore specified, it shall issue to the applicant a permit. Such permit may be conditioned upon the applicant's amending the proposal to take whatever measures are necessary to protect the public interest. The Agency shall act upon an application for permit within 90 days of the application being filed.

(g) In the event a majority of the members of the Agency determine that a permit should be denied, and any applicant who is aggrieved or adversely affected thereby shall have the right to appeal as provided in subparagraph (j) of this section.

(h) In the event any member of the Agency determines that a conditional permit should be issued the member of the Agency making such determination shall notify the other members of the Agency in writing of the conditions and reasons therefore, and the Agency shall have an additional 15 days to act with regard to the application. Should a majority of the members of the Agency agree that such permit should be conditional, the permit shall be issued on such conditions or as a majority of the representatives of the Agency shall directs. If less than a majority agrees that such permit should be conditional, the permit shall be issued without such conditions. Any applicant who is aggrieved or adversely affected thereby shall have the right to appeal as provided in subparagraph (j) of this section.

(i) No permit shall be issued unless the proposed change of use of the area shall be completed within two (2) years next after the date of the issuance of such permit. Such time may be extended for good cause upon showing that all due efforts and diligence toward the completion of the work have been made. Any permit may be revoked for non-compliance with or for violation of its terms after written notice of intention to do so has been furnished to the holder thereof.

(j) Any person who is aggrieved or adversely affected by any final order or action of the Agency shall have the right to a hearing and such hearing shall be conducted pursuant to the Georgia Administrative Procedure Act (Ga. Laws 1964, p. 338 et. reg.) as now or hereafter amended.

Section 6. The State Game and Fish Commission, through its officers and wildlife rangers, shall in addition to their other duties prescribed by law make reasonable inspections of the marshlands to ascertain whether the requirements of this Act and rules, regulations and permits promulgated or issued hereunder are being faithfully complied with. Any violations shall be immediately reported to the Coastal Marshlands Protection Agency.

Section 7. The Superior Court of the County in which the land or any part thereof lies shall have jurisdiction to restrain a violation of this Act at the suit of any person. In the event the land lies in more than one county and is divided equally between two or more counties, jurisdiction shall be in the Superior Court of any county in which said land lies.

Section 8. Posting of Permits—A certified copy of every permit issued to an applicant shall be prominently displayed within the area of the proposed activity. If the Agency deems it advisable, the applicant may be required to cause a sign to be erected bearing the permit number, date of issuance, name of applicant and such other information as the Agency may reasonably require. The type and size of the sign shall be specified by the Agency.

Section 9. In the event of sale, lease, rental or other conveyance by an applicant to whom a permit is issued, such permit shall be continued in force in favor of the new owner, lessee, tenant or other assignee so long as there in no change in the use of the land as set forth in the original application.

Section 10. Exceptions.—The provisions of this Act shall not apply to the following:
    (a) Activities of the State Highway Department incident to constructing, repairing, and maintaining a public road system in Georgia.
    (b) Agencies of the United States charged by law with the responsibility of keeping the rivers and harbors of this State open for nav-

igation, and agencies of this State charged by now existing law with the responsibility of keeping the rivers and harbors of this State open for navigation, including areas for utilization for spoilage designated by such agencies.

(c) Activities of public utility companies regulated by the Public Service Commission incident to constructing, erecting, repairing, and maintaining utility lines for the transmission of gas, electricity, or telephone messages

(d) Activities of companies in constructing, erecting, repairing, and maintaining railroad lines and bridges

(e) Activities of political subdivisions incident to constructing, repairing, and maintaining pipelines for the transport of water and sewerage; and

(f) The building of private docks on pilings, the walkways of which are above the marsh grass not obstructing the tidal flow, by the owners of residences located on highland adjoining such docks.

Section 12. Emergency powers—In the event of an emergency whether created by the Act of God, actions of domestic or foreign enemies, or in circumstances where grave peril to human life or welfare exists, the provisions of this Act shall be suspended for such period. The burden of proof shall be upon the persons or person relying upon this Section to establish that such an emergency did indeed exist.

Section 13. In the event any section, subsection, sentence, clauses, or phrase of this Act shall be declared or adjusted invalid or unconstitutional, such adjudication shall in no manner affect the other sections, subsections, sentences, clauses, or phrases in this Act, which shall remain of full force and effect, as if the section, subsection, sentence, phrase, or clause so declared are adjudged invalid or unconstitutional were not originally a part thereof. The General Assembly hereby declares that it would have passed the remaining parts of this Act if it had known such part or parts hereof would be declared or adjudged invalid or unconstitutional.

Section 14. All laws and parts of laws in conflict with this Act are hereby repealed.

About the Author

Native Georgian Reid Walker Harris (1930–2010) attended public schools in Brunswick, graduating from Glynn Academy in 1948. He earned a BA degree in political science at the University of North Carolina, Chapel Hill in 1952, and graduated from the U.S. Army's Russian Language School at the Presidio of Monterey, California. Harris earned a Doctor of Laws degree in 1958 from Emory University in Atlanta. He married Doris Nelms and had three sons—Reid Jr., Michael, and Douglas, as well as seven grandchildren. Harris was elected to the General Assembly of Georgia in 1964 and served for six years. During his terms of service, he led the floor fight to change the archaic methods of trial practice in all State Courts of Record to a system approximating the practice in federal courts. Also a key member of the Judiciary Committee, Harris helped pass a new appellant practice act relevant to all appeals to the Georgia Court of Appeals and the Supreme Court of Georgia. During this period, he served on subcommittees of the House Judiciary Committee, which revised the Georgia Criminal Code and Corporate Code.

Harris was the principal author of several significant bills, including the Georgia Surface Mining Land Use Act of 1968 and the Coastal Marshlands Protection Act of 1970, which protects five hundred thousand acres of the world's most productive estuaries.

Immediately after the passage of the Coastal Marshlands Protection Act, several who were against it threatened to bring suits to contest its constitutionality. No litigation was ever brought. There were some who indicated that they would pursue a path to seek refunds of taxes paid on marsh after the attorney general's ruling regarding marsh ownership. Since the amounts paid in had been so small, the returns on marsh being for no less than one dollar or more than five dollars an acre a year, no one pursued a remedy.

In the early summer of 1970, Harris was employed by a group of homeowners in East Beach Subdivision on St. Simons Island who had heard that a subdivision to be located between their houses and the ocean was in the planning stages. There were covenants against building on the accreted land. The homeowners also felt that a dedication had been made of the land for their use. Litigation began and lasted for many years, requiring repeated appearances before the Georgia Supreme Court. The case was finally resolved in favor of the homeowners.

CPSIA information can be obtained
at www.ICGtesting.com
Printed in the USA
LVHW091700180120
644121LV00002B/183

9 780820 356730